# DYES
from
# American Native Plants

# DYES from American NATIVE PLANTS

## A PRACTICAL GUIDE

**Lynne Richards**
**&**
**Ronald J. Tyrl**

TIMBER PRESS
*Portland · Cambridge*

Frontispiece: A home dyer from Colonial Williamsburg stirs a kettle of dye in her garden. Behind her, newly colored yarn skeins are hung to dry. Photo courtesy of Colonial Williamsburg, Williamsburg, Virginia.

All photographs by the authors unless otherwise stated.



Published in 2005 by

Timber Press, Inc.
The Haseltine Building
133 S.W. Second Avenue, Suite 450
Portland, Oregon 97204-3527, U.S.A.

Timber Press
2 Station Road
Swavesey
Cambridge CB4 5QJ, U.K.

www.timberpress.com

Printed in China
Designed by Susan Applegate

Library of Congress Cataloging-in-Publication Data

Richards, Lynne.
Dyes from American native plants: a practical guide / Lynne Richards and Ronald J. Tyrl.
p. cm.
Includes bibliographical references and index.
ISBN 0-88192-668-X (hardcover)
1. Dyes and dyeing—United States. 2. Dye plants—United States. I. Tyrl, Ronald J. II. Title.
TP897.R53 2005
667′.26′097—dc22 2004004791

A catalog record for this book is also available from the British Library.

# Contents

# Preface

During the 1930s, the Works Progress Administration sponsored the Oklahoma Indian and Pioneer Project, one of many federally funded programs that provided employment to Americans during the Great Depression. This project involved the collection of historic reminiscences from people who had lived in Oklahoma before statehood, when the region comprised the Oklahoma and Indian territories. These memories touched on all aspects of the individuals' lives. Essentially verbatim transcripts of the interviews were subsequently published within the approximately 46,000-page, 116-volume *Indian-Pioneer Papers*.

During our perusal of the *Papers*, we found numerous references to the natural dyes used to color clothing and textiles during the territorial period. These reminiscences, which we collected and published in "Folk Dyeing with Natural Materials in Oklahoma's Indian Territory" (Richards 1994), marked the beginning of *Dyes from American Native Plants*. We decided to test the botanical dye materials identified in the *Papers*, both to determine the accuracy of the memories concerning the colors produced and to experiment with various forms of dye processing.

Several plant species mentioned in the *Papers* as dye sources were referred to by unfamiliar or ambiguous common names. This led us to test additional species in hopes of discovering the elusive identities of those sources of dye. In doing so we became so enthralled with the diversity of colors obtained, and with their relationship to different plant groups, that we ultimately expanded the project to encompass many additional plants native to the south-central portion of the United States.

The preliminary results of this extensive investigation have been shared with other dye and plant enthusiasts at professional conferences and have served as a basis for Elderhostel and youth education programs. It was the encouragement we received from the participants of these conferences that motivated us to produce this comprehensive compilation of the results of our search for natural dyes.

In this text we discuss the science of color and dyeing, the history of natural dyes, dye equipment and processing, and the colors obtained from 158 native North American plant species. For obvious reasons, only a small fraction of the plant species present in North America were examined. The selected species were dissected and their parts processed in two ways and tested with five mordants. The result is a collection of more than 4,600 dye samples.

It should be remembered that natural dyeing is not an exact science. Environmental factors such as temperature, rainfall, and soil pH influence the colors produced by any dye plant from place to place and year to year. Therefore the colors that we have identified should be considered approximate representations of the hues you will realize in your own dyeing. Surprise, however, is a major part of the pleasure of natural dyeing. We hope that you will experience as much enjoyment when pursuing your natural dye projects as we have with ours.

This book consists of four parts. Chapters 1 through 4 present an overview of the science and history of dyeing. Chapters 5 through 10 delineate the colors produced by the 158 native plant species, with hue serving as the basis for organization. Chapter 11 identifies those experimental conditions that resulted in little or no dye color. Chapter 12 includes descriptions of all 158 species so as to assist you in locating these plants in your own area. For practical purposes, nonmetric (U.S.) measurements are provided for the majority of the book; metric measurements are used in chapter 12, however, as is customary in plant descriptions. Conversion tables are located at the back of the book.

The results of our dye experiments are reported in tables within Chapters 5 through 11, using descriptive color terminology. These dye results also were analyzed using the Munsell color system, which codifies color more specifically in terms of hue, value, and saturation. At the beginning of each chapter, the Munsell codes representing the results of our dye experiments are associated with the various descriptive color terminologies found within the tables, along with sample color chips that approximate the Munsell color codes. Be aware that because the ink-on-paper printing technology used in the manufacturing of this book employs a color identification system different from Munsell, these color chips are approximate, not exact, representations of Munsell color designations.

As we conducted research and wrote this book, a number of individuals provided their time, expertise, and words of encouragement. We gratefully acknowledge these contributions. Thanks are expressed to Ghisleli Ramirez-Tate and Cathy Starr, who served as laboratory assistants and monitored many steaming pots of mordants and dye, and to Paula Shryock and Crystal Small, who tirelessly assisted with many odious, tedious editorial tasks. The Oklahoma State University College of Arts and Sciences Summer Travel Program provided financial assistance, the College of Human Environmental Sciences at Oklahoma State University provided laboratory space, and the Royal Botanic Gardens, Kew, was kind enough to provide access to their library and herbarium collection. We are also grateful to Flora Oklahoma for granting us permission to use a portion of their glossary.

Special thanks are due the following individuals, who granted us use of their striking photographs: Terrence G. Bidwell, Paul Buck, Gerald D. Carr, Andrew Crosthwaite, George M. Diggs Jr., Wayne J. Elisens, Robert J. George, Steven K. Goldsmith, Bruce W. Hoagland, Charles S. Lewallen, Ronald E. Masters, Robert J. O'Kennon, Adam K. Ryburn, David G. Smith, and Dan Tenaglia. Thanks are also due the Botanical Research Institute of Texas, the Samuel Roberts Noble Foundation, and Colonial Williamsburg for permission to use their photographs. Gina Crowder Levesque was kind enough to share her fabrics dyed with native plants. Her generosity is gratefully acknowledged.

The individuals responsible for the transition of manuscript to finished book must also be recognized. Our thanks to Neal Maillet, executive editor at Timber Press, for his ability to understand our vision of how dye plants and their colors might be related in a formal book. A special thanks to Mindy Fitch, our patient editor, whose careful work ensured accuracy, consistency, and clarity throughout the book.

# 1

# An Introduction to Natural Color

Natural dyeing is the extraction of color-producing agents, called chromophores, from objects in nature and the application of these coloring agents to a desired material. Historically, natural dyes have been used to color such things as textiles, baskets, leathers, and even human bodies. Plants are the source of most natural dyes. Different parts of any plant—flowers, leaves, fruit, or bark, for example—may produce different dye colors. Some insects and shellfish have also been used to produce natural dyes.

Most modern commercial textile products are colored with synthetic—that is, human-made—dyes. These dyes, which are chemically formulated in a laboratory, provide predictable and durable colors, characteristics that are important in the mass production of fabric. In contrast, natural dyes have greater color variance, because many uncontrollable factors such as the climate or the mineral content of the soil can influence the hue, value, and saturation produced by any one dye source (Buchanan 1995). Natural dyes can also be more delicate in terms of lightfastness and washability than many synthetic dyes.

Most manufacturers of mass-produced textiles value the uniformity and durability of synthetic dyes. Textile artists, on the other hand, appreciate the more unique colors obtained from natural dyes. Natural hues often have more subtle depth or less uniform flatness in color. In *Natural Dyes* (1950, 77), Sallie Kierstead observed that "the lustrous tones of natural dyes . . . seem to be alive and to hold the light in a way that no synthetic dyes can; as they age they mellow and blend . . . taking on new richness." The appeal of naturally dyed fabrics is evident in the high prices paid for traditional Oriental rugs and Navajo blankets, which are woven with naturally dyed yarns.

Contemporary environmental concerns have stimulated public interest in dyes that produce less toxic contamination. Dyes made from naturally occurring materials tend to be less hazardous than synthetic dyes. A few large-scale textile manufacturers have begun to offer naturally dyed fabrics as alternatives to textiles colored with human-made chemical dyes. These fabrics have been used to produce items such as bed and bath linens, intimate apparel, and sportswear (Kadolph 1997).

The growing interest in natural dyes has supported the commercial cultivation and collection of natural dyestuffs within developing regions of the world, where dye production offers economic opportunities within rural communities. The government of Pakistan, for example, has encouraged the production of natural dyes by establishing a national center for dye research and information. Similarly, marketing centers for natural dyes have been formally established in Bangladesh (Green 1995; Kadolph 1997).

Advocates of natural dyeing tend to be the kind of people who appreciate a direct connection between humans and nature. In *A Handbook of Dyes from Natural Materials* (1981, x), Anne Bliss attempts to explain her own preference for natural dye over "a packet of dye powder":

> Perhaps it's because I love soil and plants . . . Perhaps it's because I don't want one of humanity's oldest arts, crafts, and trades to disappear. Perhaps it's because I can wander through the woods or the vacant lots in a town, gather up a few unsuspecting plants, take them home, and brew a lovely dye all by myself. Perhaps it's because there are sometimes surprising results in the dyepot that make me wonder and appreciate the fact that I can't always control what nature has produced. Perhaps it's because dyeing with nature feels to me like the way things are meant to be.

## The Nature of Color

Color is a natural phenomenon, the perception of which requires light. Visible light is just one type of electromagnetic radiation pulsing throughout the earth's atmosphere. Different forms of electromagnetic radiation have different wavelengths. The electromagnetic rays used for radio transmissions have the longest wavelengths, whereas X-rays have very short wavelengths. The visible light rays that produce color have moderate wavelengths.

Sir Isaac Newton discovered in the seventeenth century that visible "white" light such as that produced by the sun is actually a mixture of diversely colored light rays. When he directed a beam of sunlight through a glass prism, he discovered that it separated into bands of many different colors. Repetition of the experiment revealed that the colors were always arranged in a particular order: red, orange, yellow, green, blue, indigo, and violet. This separation and arrangement of individual colors is known as the color spectrum.

These differently colored light rays have different wavelengths. Red wavelengths are longer, violet wavelengths shorter. The dense material of Newton's prism bent these varying wavelengths at differing angles, thereby separating the white light into a visible colored spectrum. White light, then, is what the eye sees when the diversely colored light rays of a full spectrum are combined.

When light strikes a surface, the surface absorbs some wavelengths and reflects others. For example, when light strikes material that the eye perceives to be yellow, that material is absorbing red, orange, green, blue, indigo, and violet wavelengths, and reflecting the yellow. If a fabric reflects all wavelengths and absorbs none, the eye perceives the material to be white, because white light encompasses the full spectrum. If the fabric reflects none of the wavelengths but absorbs all of them, the fabric is perceived to be black, which is the absence of reflected wavelengths.

For a surface to reflect a particular color, wavelengths of that color must be present in the atmosphere and striking the surface. Sunlight and full-spectrum electric lights most accurately represent the color-reflection properties of a surface, because these light sources contain the full range of wavelengths. On the other hand, a light source that lacks wavelengths of a specific hue undermines reflection of that hue by surfaces that the light illuminates. For example, if a light low in blue rays hits a surface dyed to ab-

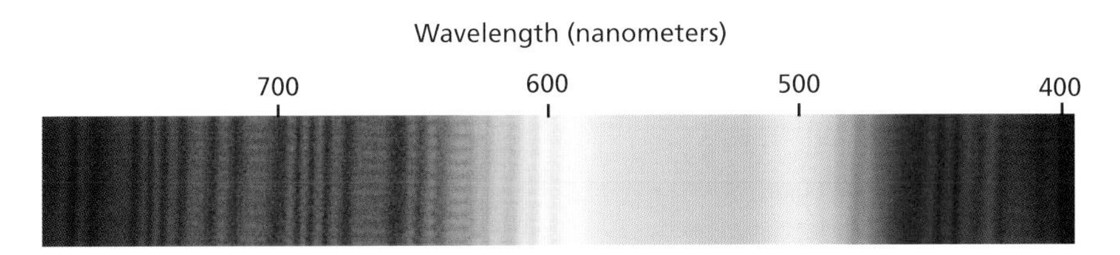


The color spectrum.

sorb all rays except blue, that surface will appear nearly black. In other words, a light source that is low in blue light rays will produce very little blue light to be reflected. The variability of artificial lighting, with respect to the spectrum of wavelengths emitted, explains why objects often appear to be different in color under different light sources. For example, clothing purchased in a store under fluorescent light may appear to be a slightly different color when viewed in sunlight or incandescent light.

During the process of natural dyeing, chromophores are extracted from a plant by soaking the plant material in water. When a textile is placed in the water, the extracted chromophores become attached to the textile fibers. These chromophores subsequently determine which wavelengths will be absorbed by the textile and which will be reflected, thereby creating the perceived fabric color.

Not all plant chromophores are equally water-soluble. Some that give color to a plant may not be released into the dyebath and thus will not be available for attachment to fibers. Also, the color-absorbing and color-reflecting properties of a chromophore may be altered during the dye process by heat, the acidity or alkalinity of the dye solution, the presence of minerals, or other factors. As a result, the colors observed in the tissues of a plant may not be the same colors that will be reflected by a textile dyed with that plant.

Chlorophyll chromophores reflect green light and thereby give most plants their typical green color. These chromophores, however, are relatively insoluble in water and are not easily extracted and affixed to the surface of textile fibers. As a result, strong green dyes from nature are rare.

Yellows, golds, and oranges, the most commonly obtained hues from natural dyes, are produced by plant carotenoid and flavonoid chromophores. Carotenoids, for example, are the chemical compounds that give carrots their typical orange color. Anthocyanins, a type of flavonoid chromophore, are responsible for most of the red, violet, and blue hues observed in plants. Anthocyanins, however, exhibit tenuous attachment to fibers and as a result usually produce weak dyes that fade easily. Anthracene chromophores, which are found in such natural dyes as madder and cochineal, produce much more durable red hues (Cannon and Cannon 1994; Tull 1987).

The unusually strong blue dye obtained from species of indigo (*Indigofera suffruticosa* or *I. tinctoria*) is produced by the chromophore indigotin. It is insoluble in water, but through a process of reduction (or vatting) in dilute alkali, it is changed to a colorless liquid that does have affinity for fibers. After a textile is dipped into the clear vat solution and removed, the indigotin chemically develops or changes upon exposure to air. Through this process of chemical oxidation, the indigotin thereafter reflects blue light rays.

The brown hues observed in bark, wood, and autumn leaves are produced by tannic acid chromophores. These chromophores are water-soluble and have a strong affinity for fibers. As a result they produce good, durable dyes. Rather than fading when exposed to light, as happens with many other natural dyes, tannic acid chromophores will actually darken (Tull 1987).

## Color Mixing

Although a full range of colors can be obtained from natural dyes, plants that give bright blues, greens, and true reds are few in number. In some cases elusive hues can be produced through a process of top-dyeing, which involves dyeing one color over another. Red, blue, and yellow are primary colors, which means that they cannot be created by mixing other hues or by top-dyeing. On the other hand, two primary colors can be combined to produce a secondary color. Green, for instance, is a secondary color that can be achieved by dyeing fibers with a yellow-reflecting chromophore and then top-dyeing with a blue-reflecting chromophore. When a textile reflects both yellow and blue light rays, the eye sees green. Other secondary colors are orange (red and yellow) and purple (red and blue).

In 1766, Moses Harris published a color wheel that illustrated these color relationships. The primary colors of red, yellow, and blue are separated on the wheel by the secondary colors of green, orange, and purple. Their placement reveals which primary colors are combined to form each secondary color. Between each primary and secondary color is a tertiary color. Tertiary colors are produced by combining a primary color and an adjacent secondary color. For example, combining primary yellow and secondary orange produces a tertiary yellow-orange. The proportions of yellow to orange dye will determine whether the reflected light is more yellow or orange.

## Value and Saturation

The lightness or darkness of a hue is referred to as its value. A color with a high value contains much white and is said to be pastel. A low value signifies that the color is nearly black.

Black can be produced by top-dyeing a textile with all three primary col-

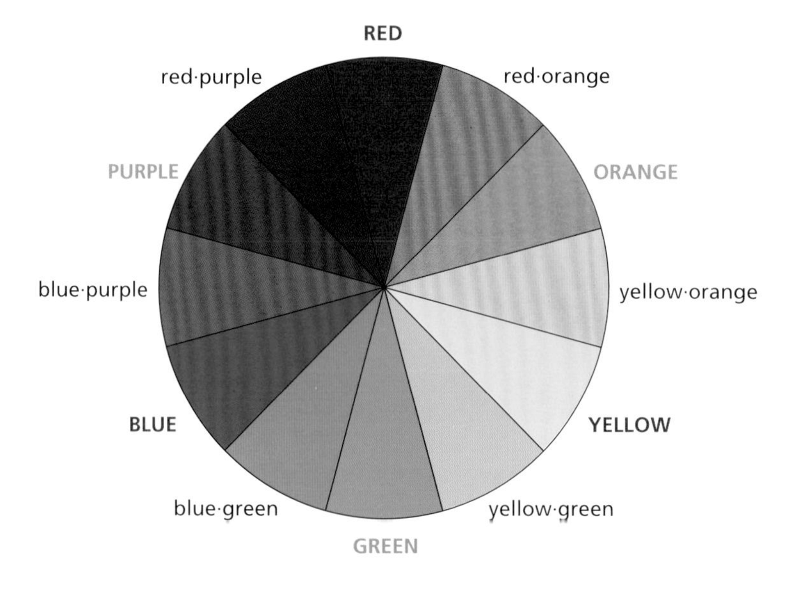


The color wheel illustrates the relationships among the three primary hues (red labels), red, yellow, and blue; the three secondary hues (green labels), orange, green, and purple; and the six tertiary hues, red-orange, yellow-orange, yellow-green, blue-green, blue-purple, and red-purple.

ors: red, yellow, and blue. When a fabric is dyed yellow it absorbs red and blue light rays. When top-dyed with red it also absorbs yellow and blue. Dyeing again with a blue dye causes absorption of yellow and red. In other words, dyeing with all three primaries cancels the light-reflection properties of each chromophore (and their secondary and tertiary hues), resulting in black. Therefore, a dye that produces a low-value (dark) hue is reflecting very little light and absorbing almost all light that strikes its surface.

A high-value white or near-white textile, in contrast, has very little light absorption. It reflects all or almost all of the colored light that strikes its surface. Remember that when the light rays of a full-color spectrum are combined, the visual result is white light. Therefore, a high-value (light) hue is reflecting almost all light rays and thus appears to be pale, pastel, or nearly white.

The saturation of a hue refers to its degree of brightness, strength, or intensity. Highly saturated colors are very strong (for example, a bright red), whereas colors with low saturation appear dull or grayed. A hue's saturation can be lowered, or grayed, by adding a little of its complementary color. A color's complement is that which is directly opposite it on the color wheel. For example, the saturation of a bright red can be lowered or dulled by top-dyeing with a light green wash, because red and green are complementary colors.

## The Munsell Color System

Formally published in 1905 by Albert Henry Munsell, a Boston artist and teacher, the Munsell color system provides a means of precisely and consistently describing colors with respect to hue, value, and saturation. Munsell established numerical scales of equal visual increments for each of these three characteristics.

In his system, any color can be defined as a point in a three-dimensional color space. Hue is expressed by using the initial letter of the color name plus a number designating the degree to which the color matches that hue on the color wheel, or has elements of an adjacent hue. For example, 5R is very red, whereas 10R is closer to the next hue, yellow-red (YR). Similarly 5YR is very yellow-red, whereas 10YR is closer to yellow (Y). Colors are abbreviated as follows:

| | |
|---|---|
| R | red |
| YR | yellow·red |
| Y | yellow |
| GY | green·yellow |
| G | green |
| BG | blue·green |
| B | blue |
| PB | purple·blue |
| P | purple |
| RP | red·purple |

Value is expressed by a number, ranging from 0 for pure black to 10 for pure white. Saturation (what Munsell called chroma) is likewise expressed by a number, ranging from 2 for a dull color that has very little hue and is very close to neutral black, gray, or white, to 20 or higher for a very bright or intense color.

The complete Munsell notation for any one color consists of representations for hue, value, and saturation, in that order. For example, a red of medium value and bright saturation might be designated as 5R 5/12, whereas a yellow-green of low value and dull saturation might be designated as 2.5GY 2/4. Presently published as *The Munsell Book of Color*, this system has been accepted as an international standard for color and is used in many disciplines.

## Mordants

Natural dye materials that produce strong, durable colors without the addition of other substances are called substantive or direct dyes. Plants high in tannic acid, for example, such as sumac (*Rhus* species) and walnut (*Juglans* species), produce direct dyes. However, not all natural dye chromophores automatically attach to textile fibers with a durable molecular bond without assistance from chemical additives. Dyes that need this assistance are called adjective or mordant dyes. Mordants are water-soluble chemicals that create a bond between dye and fiber. Although only adjective dyes require the presence of a mordant for dyeing to occur, both adjective and direct dyes can benefit from the use of mordants in terms of color richness and durability.

In many cases a mordant will also influence a plant chromophore's properties of light absorption and reflection. Using yarns treated with different mordants, therefore, increases the diversity of hues that can be obtained from any one dye source. When dye assistance is to be provided by a mordant, the textile may be subjected to a mordant presoak, or the mordant may be added to the dyebath, or the dyed material may be immersed in a final mordant rinse.

## Environmentally Responsible Dyeing

The dye recipes used by pioneer settlers often suggested collecting as much as a peck (or a quarter of a bushel) of plant material before beginning the dye process. Historically, of course, wild plants were abundantly and readily available. Therefore, the harvesting and subsequent discarding of unneeded surplus plants was not perceived to be wasteful.

Today we live in a world characterized by diminishing wilderness, endangered species, and environmental pollution. We each must accept responsibility for protecting and conserving our natural resources. Nature provides us with a bountiful array of dye materials from which to select. In the spirit of appreciation, it is important to collect no more than is needed and to harvest only those dye plants that are growing abundantly. Leaving roots intact to encourage regrowth, restricting the quantity of fruit and flowers collected to facilitate reseeding, and trimming wood and bark only from fallen branches and logs will help preserve dye plants for enjoyment by future generations.

# A Brief History of Natural Dyeing

The origins of natural dyeing are hidden in the mists of prehistory. No doubt early humans observed the staining properties of plants when they plucked fruits or flowers and noted the various colors of rocks and soils. Human bones in prehistoric Neolithic graves have been found powdered with colored mineral pigments, suggesting that these people used the colors of the earth to add decoration to their clothing and bodies.

To these early humans, the extraction and transfer of colors from nature to objects of their choice must have seemed magical. As a result, superstitions evolved concerning the ritualistic procedures whereby the wonders of natural dyeing were accomplished, and certain colors were believed to have magical powers (Knaggs 1992). Even among some twentieth-century cultures, superstition continued to be a part of the dyeing process. On the island of Rotti, in Indonesia, members of the indigenous culture traditionally believed that evil spirits enjoyed dipping their hands into dyepots to deprive the dyes of their effectiveness. Charms constructed from the wood of the lontar tree and hen's feathers were hung above the dyepots to ward off such mischievous evil (Buhler 1948).

## Natural Dyeing in Antiquity

Evidence of natural dyeing in antiquity has been discovered in many parts of the world (Lillie 1979). Natural dyed fabrics were commercially produced in China as early as 3000 B.C., according to an ancient Chinese document (Adrosko 1971). Textile fragments dyed with roots from the madder plant (*Rubia tinctoria*), which produces red, were found at Mohenjo-Daro, an archaeological site in Pakistan dating to around 2500 B.C. Similar fabrics have also been unearthed in Egyptian tombs (Dean 1999). Ancient Hebrew women collected shield lice from the branches of oak trees to produce a red dye, a color source now known as kermes, and the Bible records the use of many other diversely colored fabrics (Sandberg 1994; Wilson 1979).

The ancient Phoenician dye industry, which was located along the eastern coast of the Mediterranean Sea as early as 1500 B.C., is credited with the discovery of several beautiful purple dyes, one of which was known as Tyrean purple. These purple hues were obtained from glandular secretions produced by a number of mollusk species (Knaggs 1992; Sandberg 1989, 1994). Dibromoindigo, the dye chromophore found in these secretions, changes from yellow to red-purple upon exposure to air (Bliss 1981; Wilson 1979). Tyrean purple was extremely expensive to produce because of the complicated vatting process it required and because as many as 12,000 mollusks were needed to produce only 3.5 ounces of dye (Held 1973). As a result of its costliness, Tyrean purple became the color of royalty. Modern archaeological excavations along the eastern Mediterranean coast have turned up huge mounds of old shells, revealing the presence of those ancient dye works.

As early as 2000 B.C., mordants were fortifying natural dyes in India (Held 1973). Pliny the Elder, a Roman chronicler of nature from the first century A.D., described the mordanting process that he observed in Egypt:

> After pressing the material, which is white at first, they saturate it, not with colors, but with mordents [*sic*] that are calculated to absorb color. This done, the cloths, still unchanged in appearance are plunged into a cauldron of boiling dye and are removed the next moment fully colored. It is a singular fact that although the dye in the pan is of one uniform color, the material when taken out of it is of various colors, according to the nature of the mordents that have been respectively applied to it: these colors will never wash out. (Haberly 1957, 159)

## Natural Dyeing in Historic Europe

As the Roman army moved northward through central Europe during the first century B.C., they encountered members of the local Celtic population whose bodies had been dyed and tattooed blue using a plant dyestuff known as woad (*Isatis tinctoria*). This European use of woad, a plant that was indigenous to southeastern Russia and lands around the eastern Mediterranean Sea, suggests that European trade in dyestuffs existed even at this early date (Knaggs 1992).

By the eighth century A.D., European trade in dyestuffs flourished. Fabrics and commercial dyes were transported along established trade routes from Asia and the Middle East westward to ports along the Italian peninsula. From there the commerce moved northward through the Alps to trade fairs. The famous fairs of the Champagne region of France, which were located halfway between Flanders in the north and Italy in the south, served as commercial clearinghouses for all of Europe, and dyes were among the most important products sold. The fairs were a major stimulus to European textile production, and communities devoted to textiles were built near important market towns. Complex systems of craft guilds were developed to regulate the training of apprentices and the production of quality textile merchandise.

Until the sixteenth century, natural dyeing in Europe was highly secretive, and dye recipes were closely guarded. Then in 1548, Gioanventura Rosetti published *The Plictho*, a reference describing the dye processes and recipes used by dyers working in the cities of Italy. This publication became a standard reference for commercial natural dyers until the end of the seventeenth century. At that time the Academie des Sciences in Paris began organized efforts to advance the science and chemistry of natural dyeing. This resulted in experimental searches for new dyestuffs (Robinson 1969). The discovery of America also contributed to this scientific era in the history of natural dyeing. New World dyestuffs such as logwood (*Haematoxylon campechianum*), fustic (*Chlorophora tinctoria*), and cochineal insects (*Dactylopius coccus*) expanded the range of hues that could be obtained from natural dyes.

## Natural Dyeing in North America

The Europeans who settled along the eastern coast of North America during the seventeenth and eighteenth centuries brought their knowledge of European dyes and dyeing processes to North America. The settlers preferred to use the dyestuffs with which they were accustomed, and so significant quantities of commercially prepared European dyes were imported into the American colonies. An accounting of "Goods and Produce imported into the several Provinces in North America" in 1770, alone, included 70 tons of imported dye woods (Sheffield 1784, table 4).

The dyestuffs imported into the American colonies were very expensive, having passed through the hands of many middlemen. After the Revolutionary War, Asa Ellis published *The Country Dyer's Assistant* (1798), the first manual concerning natural dyeing in North America. In it he lamented:

> For a great proportion of the ingredients employed in dyes, we depend on Europe to furnish. . . . As we attempt an independence of their markets, they increase their duties on dyestuffs which we import. Not one cask, of Cochineal, can we obtain from our sister continent, South-America; from thence it must pop through the hands of Spain and England. From England we receive it, at an extravagant price. . . . Foreign nations receive a large revenue from this country, for the dyestuffs we import. Does it become an independent nation, to be thus dependent on others, for articles, which, perhaps, may abound in our own country? (137–139)

Because of the high cost of commercial dyestuffs, and because many rural early Americans did not have immediate access to trading centers, some settlers planted the seeds of familiar dyestuffs in their gardens to ensure the availability of coloring materials. Woad (*Isatis tinctoria*) for blue was a popular garden dye plant, although it exhausted the soil and thereby necessitated the constant opening of new tracts of land (Stearns 1964).

European settlers also experimented with the unfamiliar native plants that they discovered within their new environment. They encountered local species related to their European dye plants, such as a North American sumac (*Rhus* species) similar to the "diar's shumach" of England. They also encountered completely new dye sources. While some individuals shared the results of these discoveries with neighbors, others did not. Reportedly, one colonial housewife in New England formulated a process for obtaining a rare pink color but refused to divulge the source of this dye and took the secret to her grave. That particular hue became known as Wyndym pink, named after the town in which she lived (Harbeson 1938).

European settlers also learned about American dye plants from resident Native Americans, who produced red from bloodroot (*Sanguinaria canadensis*), green from algae, and yellow from lichens. The poisonous fruit of pokeweed (*Phytolacca americana*), an indigenous North American herb, was favored by Native Americans as a colorant for baskets (Weigle 1974). The American colonists observed, however, that many of the hues produced by Native Americans quickly faded, suggesting that mordants may not have been used.

Natural dyes remained the only alternative for dyeing textiles until 1856 when an English scientist, William Perkin, successfully created in his laboratory a mauve dye from coal tar (Adrosko 1971). Other human-made dyes soon followed, and by the 1870s synthetic dyes were commercially available to home dyers in the United States (Nast 1981). However, as American pioneers moved westward into largely unpopulated regions of the central frontier, commercial outlets for synthetic dyes were often slow to follow. Therefore, to people residing great distances from commercial centers, natural dyes remained important throughout the nineteenth century (Richards 1992, 1994).

Emeline Crumb, a Kansas pioneer, recalled her family's use of natural dyes for coloring homespun wools and unbleached muslin:

> The native plants and barks were used. . . . Good browns were produced with walnut bark, when properly set. . . . A good dark slate or grey was secured by walnut bark, set with sumac 'bobs'. . . . Golden Rod was used for yellows—the tint called 'Nankeen.' Rusted iron—or iron filings—set some colors. . . . A bolt of strong unbleached muslin, after passing thru the dyes, made quite nifty dresses. . . . The Indians used many kinds of roots, barks and berries, making lasting colors. . . . But they were not inclined to impart their secrets to the Pale Face. (Stratton 1981, 67–68)

Just south of Kansas was Indian Territory, later to become the state of Oklahoma. Following passage of the Indian Removal Act in 1830, more than thirty different Native American tribes were resettled on this land north of the Texas border. Those that came from the southeastern United States were often accompanied by their African American slaves. After the American Civil War, European Americans also migrated into the territory. All three of these ethnic groups used natural dyes to color both apparel and household fabrics. Ransom Parris, for example, recalled obtaining a red dye from "pacoon roots" (Works Progress Administration 1937, 69: 352). He was probably referring to bloodroot (*Sanguinaria canadensis*), also known as red puccoon, a popular dye among Cherokees (Hamel and Chiltoskey 1975). Bloodroot grew in the deciduous forests of the eastern part of Indian Territory.

In the semiarid southwestern part of Indian Territory, prairie mesquite (*Prosopis glandulosa*) was a common shrub. Martha Martin, who lived in that region in 1886, recalled how "the mesquite roots taken green and boiled with real strong iron and copper" made "a pretty golden colored dye" (Works Progress Administration 1937, 60: 500).

The bois d'arc tree, or Osage orange (*Maclura pomifera*), was another popular source of yellow dye among early residents of the southern plains. Minnie Parks, whose uncle worked in a Texas bakery, recalled that "he used to send mother great bundles of flour sacks. Mother made all of our clothes out of them. . . . She would use bois d'arc bark to dye the sacks bright yellow for aprons and quilt tops" (Works Progress Administration 1937, 69: 284).

If a pioneer family lived within traveling distance of a trading center, they sometimes purchased commercially prepared natural dyes. John Harrison, an African American, recalled that "indigo was purchased at trading posts and all shades of blue could be made" (Works Progress Administration 1937, 39: 325). On the other hand, Elijah Culberson remembered that his Cherokee family used "wild indigo," probably either blue wild-indigo (*Baptisia australis*) or yellow wild-indigo (*B. tinctoria*) (Works Progress Administration 1937, 22: 218).

Sarah Harland, a Choctaw, met her needs for dye by growing indigo in her garden during the American Civil War:

> A druggist in Bonham [Texas] gave me some indigo seed. . . . I planted the seed. When it was just blooming, the old negro man . . . cut it down, put it in barrels, and pounded it . . . then he said let it rot, so we did and I tell you it beat any . . . smell I ever smelled. Then he put it under a press and pressed the juice out, strained it, put it in a boiler and boiled it for two days, from 10 gallons down to one, until it was thick like syrup, then put it in dishes and set it in the sun to dry; it evaporated and became hard. We tried it in water, and to our great joy found it just what we wanted. (Works Progress Administration 1937, 106: 153–154)

The seeds that Sarah received may have been those of *Indigofera suffruticosa*, a species of indigo that was commercially cultivated in the southeastern United States during the eighteenth century (Adrosko 1971).

Some pioneers combined their natural dyes with tie-dye techniques to produce patterned textiles. Levi Ketcher described his Cherokee grandmother's procedure: "Sometimes you take white [corn] shuck, tie it in places around the hanks [of yarn] and [after dyeing] this leaves white spots in [the] thread" (Works Progress Administration 1937, 50: 438). Mattie Huffman, who lived in Kansas, remembered her aunt's technique:

> [She] would gather quite a number of corn shucks, cut them in two cross-wise and sit down with her lap full of them and a

> quantity of yarn. Then she would begin winding the knitting yarn [around] . . . these shucks; about 12 inches on each one, and leaving about as much yarn between each two. When the entire skein was so prepared on the shucks it was then put into the dye. When it was taken out, if the dye was red, for instance, next to the shuck would be the natural white color of the yarn, next to that a pink shade, and . . . on the outside it would be red. . . . They called [this] clouded yarn. (Stratton 1981, 67)

As commerce grew in central portions of the United States, however, reliance upon natural dyes declined. Commercial dyes, as well as predyed fabrics, reduced the need for laboriously collecting materials from nature and chopping, soaking, and boiling those materials to acquire color. It wasn't until the arts and crafts movements of the early 1900s and 1960s that renewed interest in natural dyeing motivated efforts to collect and publish the traditional dye recipes used by past generations. Today, textile artists and textile historians are continuing this process of experimentation and documentation, verifying the natural dyes used by our ancestors, in addition to discovering new environmental sources of color.

# 3

# Dye Supplies and Equipment

The beauty of natural dyeing has to do not only with the quality of the colors obtained but also with the ease and economy with which the process can be performed. Natural dyeing does not require the use of elaborate or expensive supplies, equipment, or facilities. In fact, most of the items needed are readily available in a typical household or grocery store. In this chapter, we describe essential supplies and equipment.

## Textile Fibers

Natural dye chromophores attach more readily to some fibers than to others. In general, protein fibers from animals produce better dye results than do cellulosic fibers obtained from plants. Wool, hair, and silk are protein fibers, whereas cotton, linen, jute, and hemp are cellulosic. Natural dyes can also be used to color leather, which is protein-based. Synthetic fibers such as polyester or acrylic usually do not bond well with natural dye chromophores.

Textile fibers must be clean before they can be dyed; otherwise, the dye results will be spotty or mottled. In their natural state, wool and hair fibers contain oils, whereas silk has a coating of gum, called sericin. Most silk yarns and fabrics are degummed before being sold commercially. However, yarns and fabrics made from wool or hair may still contain their natural oils, or they may have been contaminated with oils and chemicals during the manufacturing process. Removing these oils and chemicals by a process of scouring is important to successful dyeing. The purchase of prescoured yarns and fabrics, on the other hand, eliminates the necessity of scouring before dyeing.

Natural dyes may be applied to textiles in the fiber, yarn, or fabric stage. Dyeing fibers requires subsequent carding and combing, after which the fibers are spun into yarns and woven, knitted, crocheted, or knotted into a fabric. Many natural dyers are not interested in conducting such extensive processing. Dyeing at the fabric stage, on the other hand, introduces the problem of obtaining smooth and even color across the entire expanse of the fabric. This is difficult to achieve without the use of very large dye containers that allow the material to freely float within the dye solution. Due to these difficulties, therefore, dyeing at the yarn stage is preferred.

## Mordants

Mordants, which are usually metallic salts, can be purchased from chemical supply companies or dye suppliers. Although metals are natural components of the human body in trace amounts, concentrated metallic salts can be toxic. Therefore mordants should be carefully handled and stored. It is safest to purchase mordants in lumps or in a granular state, when these options are available, because fine powder can become airborne and inhaled. Store mordants away from light in tightly sealed glass or plastic containers or, when small children are present, in a locked cabinet or case. When conducting dyeing experiments with children, it is best to use yarns that have been premordanted, thereby eliminating contact with these chemicals (Turner 1992).

The mordants most commonly used with natural dyes are alum (potassium aluminum sulfate), iron or copperas (ferrous sulfate), copper or blue vitriol (copper sulfate), tin (stannous chloride), and chrome (potassium dichromate). However, due to the lethal toxicity of chrome, both to the environment and the dyer, this mordant should not be used for dyeing in the home or in facilities not equipped with a ventilation hood and means for disposing of hazardous materials (Van Stralen 1993). None of the dye recipes included in this book require the use of a chrome mordant.

Alum, when used as a mordant, usually brightens the colors obtained from a dye source. Tin also produces bright colors and is especially useful for creating more intense or saturated yellows, oranges, and reds. Copper, on the other hand, improves the likelihood of obtaining a green hue. Iron usually darkens or saddens hues, producing grays, browns, blacks, or olive-greens. Yarns dyed with copper are often most resistant to fading when exposed to light. In contrast, tin and iron are somewhat less effective in regards to improving the lightfastness of a dye, but they usually give greater fastness than can be achieved without the use of a mordant.

Most mordant recipes call for the addition of cream of tartar or tartaric acid (potassium hydrogen tartrate), which can be found in the spice section of any grocery store. Cream of tartar reduces fiber stiffness or brittleness, which can occur as a result of mordanting. It also serves to brighten colors, especially yellows and reds.

Tannic acid (tannin) is found naturally in many direct dyes and can be used as an additive agent when working with adjective dyes. It is identified by some dyers as the best additive to use when dyeing cellulosic fibers such as cotton or linen (Dean 1999; Tull 1987). Tannic acid, which is available from chemical supply companies, tends to darken colors, and this darkening may intensify with prolonged exposure to light.

For more information on mordants, see chapter 1.

## Water

Dyeing with natural materials usually involves soaking or heating the dyestuff in water so as to release water-soluble chromophores and thereby create a dyebath. Textiles are then submerged in the dyebath, whereupon the chromophores bond with the fibers. Thereafter the textile is rinsed in water to remove excess dye. Water, then, is an essential component in the natural dyeing process.

Tap water from a well or municipal water works usually contains minerals that will influence the colors obtained during natural dyeing. For example, water that is rich in iron will darken dye results. Dyeing with distilled or deionized water eliminates this potential problem.

## Disposal

Just as important as the successful bonding of dye and fabric is the proper disposal of the mordant and dye solutions to protect both the dyer and the environment. Although mordants and natural dyestuffs are materials extracted from nature and are less hazardous than many synthetic dyes, these materials should be treated with caution when in concentrated solutions. As noted previously, chrome should not be used because of its toxicity.

Alum, iron, copper, tin, and tannic acid are generally used in such low concentrations that they do not exceed the limits set by public-owned treatment works (POTWs). Therefore, dyers whose homes or workshops are connected to a city sewer system usually dispose of mordant and dye liquids by pouring them into a toilet or sink. Caution! To avoid contamination of food or cooking equipment, do not pour these liquids into a kitchen sink. Also, natural dyes and their mordants can stain porcelain if allowed to stand. After disposal of the solutions, immediately flush the toilet several times or rinse the sink repeatedly.

If the toilet or sink are connected to a septic system, mordant solutions should not be poured into them because they can cause the system to malfunction. Instead, the solutions can be diluted two- or three-fold with water and poured on the ground in areas away from food and garden plants, septic systems, wells, and the play areas of children and pets (Bliss 1981). In addition, the dumping area should be immediately and thoroughly soaked with water from a garden hose to reduce possible damage to surrounding plants. Do not hesitate to contact personnel at the local water treatment facility or state agencies dealing with environmental quality when large quantities of mordant liquid must be disposed of or if you have questions about the appropriateness of a disposal method.

Chemical suppliers in the United States are required to provide, upon request, a Material Safety Data Sheet (MSDS) for each chemical purchased by a customer. This data sheet provides specific recommendations for how to handle, store, and dispose of the chemical. Be sure to request one when purchasing mordants.

## Equipment

Most natural dyes are processed by either heating the dyestuff in order to release the chromophores (hot water process) or by decomposing the plant material in water (tepid water process). These two processes require different kinds of equipment.

Historically, some natural dyers employed dye kettles made from assorted metals as substitutes for mordants. Heating a dyestuff in an iron pot, for example, somewhat replicated the use of iron as a mordant. Today natural dyers use enamel pans to prevent the composition of their equipment from distorting dye colors. If the enamel on a dyepot becomes chipped or cracked, it should be repaired with enamel appliance paint. Otherwise the pan's base metal may leach into the dyebath and alter the color. Utensils for lifting materials in and out of mordant and dye solutions should also be made of nonreactive materials. Spoons molded from heat-resistant plastic or made of enameled metal are a good choice.

Dyepots should be large enough to allow the textile to be completely immersed and swirled around in the dyebath. When dyeing only small pieces of yarn or fabric as test samples, small enamel pans will suffice. For larger quantities of material, use a 4- to 5-gallon pot for each pound of wool (dry weight) to be dyed.

Ideally a home dyer will have access to a well-ventilated workroom in which mordants and dyes can be processed. This room should be equipped with a small stove (a portable camp stove will suffice), sink, worktable, and drying rack or surface. When available, a vented hood or fan should be used to disperse the fumes. In good weather a dye workshop can be set up outdoors, or, if necessary, mordant solutions and dyebaths can be heated on the stove in a well-ventilated kitchen. When dyeing with children, a slow cooker or Crock-Pot may be a safer alternative to a stove.

Mordants and dyestuffs can be lethal if consumed by pets or humans. If a kitchen is used during the dyeing process, avoid any possibility of contamination by making sure dyes and food are never prepared at the same time. Also, do not dispose of mordant and dye solutions by pouring them into a kitchen sink that is used for food preparation, and never use dye pans or other equipment for cooking food, no matter how thoroughly they have been washed.

When dyeing involves the decomposition or tepid water process, a dyestuff is soaked in room-temperature water for an extended period of time without heating. Glass canning jars work well for this, as do large plastic tubs with lids. When using canning jars, small plastic bags should be placed over the top of the jar before screwing on the metal lid. This prevents the dyestuff from coming into contact with and reacting to the metal of the jar lid.

Several other pieces of equipment are needed for natural dyeing, regardless of whether the water to be used will be hot or tepid. Rubber gloves are an absolute necessity. Plant dyestuffs, mordants, dye solutions, and newly mordanted or dyed textiles should not be handled without them. In addition, an accurate scale is useful for measuring mordants and calculating the appropriate amounts in relation to the weight of the textile to be treated. A large (say, 4-cup) glass or plastic measuring cup can be used to measure dyestuff and dye liquid, prewet small amounts of the textile, and rinse dye test samples. Big tubs will be required for prewetting and rinsing fabrics or yarns if large quantities of material are to be dyed. A large strainer is needed to extract vegetable matter from the dyebath, while a stout scrubbing sponge is useful for removing dyestuff residue from the insides of jars and pans. When all of these items have been assembled, the dyeing can begin!

# The Processes of Natural Dyeing

Protein fibers such as wool and silk are most receptive to dyes that are slightly acidic, whereas cellulosic fibers such as cotton have greater affinity for alkaline dyes. Acidity and alkalinity refer to the hydrogen ion activity (concentration) of a solution expressed as pH. Values range from pH = 0 (very acidic) to pH = 14 (very alkaline), with 7 as a neutral point.

Most natural dyes are slightly acidic. As a result, they bond best with protein fibers. Because wool is the fiber most frequently used with natural dyes, all of the procedures and dye recipes provided in this book pertain to it. Also, the dye instructions are written from the perspective of working with wool yarns. Realize, however, that these same dyes can be used to dye silk, or wool fibers and fabrics, if desired. They also may provide some color to cotton or linen.

During the process of dyeing, wool must be treated gently. Extreme changes in temperature, a very high temperature, or vigorous movement can cause wool to mat, felt, shrink, and become harsh and brittle. When heating a solution containing wool, therefore, increase the temperature gradually and bring it only to the point of steaming. Never boil wool. Periodically swirl the wool gently around in the mordant or dye liquid, but do not stir vigorously or frequently. Allow a solution containing the wool to gradually cool to room temperature before removing the textile from the liquid to avoid shocking the material with a rapid change in temperature.

## Forming Yarn Skeins

When wool is submerged in a liquid for the purpose of scouring, mordanting, or dyeing, it must be able to float freely in the solution. However, when working with yarns it is also important to prevent tangling. These two goals can be accomplished by loosely tying yarns into skeins before treatment. To create a skein, extend your arm and bend it at the elbow so that your fingers point toward the ceiling. Extend your thumb away from your hand to produce a U-shaped notch. Wind the yarns with a circular motion from the notch on your hand, down and under your elbow, and back to the notch. Repeat this as often as needed to produce the skein.

Carefully slip the skein of yarn off of your hand and elbow. Securely tie the two ends of the yarn strand together. Then, using separate pieces of string, loosely tie one or more circles around the yarn to keep the yarns aligned and prevent tangling. These ties must not compress the yarn. If secured too tightly they will produce white spots on the yarn when it is dyed. Using cotton string to secure wool skeins will aid in finding and clipping those ties at the end of the dyeing process, because cotton will absorb less dye than wool and will therefore be a slightly different color.

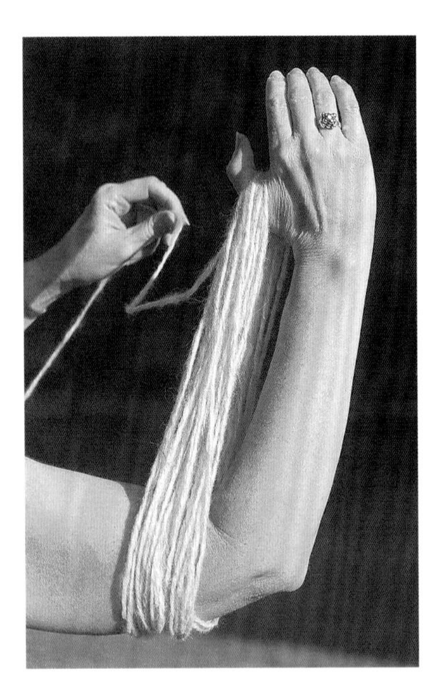

Left: Skeins are formed by winding yarn from the hand to the bent elbow and back again.

Below: Loosely tied skeins are unlikely to tangle during the dye process.

## Scouring

Scouring is a process whereby dirt and oils are removed from the wool fiber. If not removed, such foreign matter will prevent the dye from bonding uniformly with the fiber, resulting in blotched or mottled color. If commercially scoured yarns and fabrics are used, however, scouring can be omitted from the dye process.

To scour wool yarns, submerge loosely tied skeins in an enamel kettle containing room-temperature distilled or deionized water, and soak for one hour. Squeeze the yarns gently to push water into them and remove air. Lift the yarn from the kettle and mix a very small amount of mild liquid detergent into the water. Replace the wool. Avoid using an excessive amount of detergent, as it will be difficult to rinse out, and do not use a detergent that contains additives such as brighteners, bluing, or bleach. Squeeze the yarns gently to push the detergent into the skeins.

Put the kettle of water, detergent, and yarn on the stove and gradually increase the temperature until steam begins to rise from the water's surface. Swirl the yarn around in the solution once or twice using a heat-resistant spoon. Once the detergent solution begins to steam, remove the pan from the heat.

Allow the wool and detergent solution to gradually cool to room temperature. Occasionally squeeze the yarns gently to move the solution into the yarns. Wear rubber gloves to protect your hands from the heat.

When the detergent solution has cooled to room temperature, remove the yarns and gently rinse them in repeated baths of distilled or deionized water. Make sure to rinse thoroughly, as residual detergent may affect the fibers' reactions to mordant and dye.

Gently squeeze out excess water. Hang the skeins and allow them to drip-dry, or spread the yarns and dry flat.

## Mordanting

A mordant may be applied as a presoak or added to the dyebath, or the textile may be submerged into a final mordant rinse. Premordanting is the procedure used most frequently by natural dyers, because large quantities of textile can be treated and stored until needed for dyeing. The amount of mordant used is based upon the weight of the wool to be treated (see table 1). It is important not to use too much mordant, because excess amounts will cause the textile to become harsh and brittle.

| MORDANT | QUANTITY | | |
|---|---|---|---|
| | OUNCES | GRAMS | TEASPOONS |
| Alum recipe | | | |
| Alum (potassium aluminum sulfate) | 4.0 | 113.8 | 27.0 |
| Cream of tartar (potassium hydrogen tartrate) | 1.0 | 28.5 | 9.0 |
| Copper recipe | | | |
| Copper (copper sulfate) | 0.5 | 13.7 | 2.5 |
| Cream of tartar (potassium hydrogen tartrate) | 0.5 | 13.7 | 5.0 |
| Iron recipe | | | |
| Iron (ferrous sulfate) | 1.0 | 28.5 | 3.0 |
| Cream of tartar (potassium hydrogen tartrate) | 0.7 | 19.0 | 7.0 |
| Tin recipe | | | |
| Tin (stannous chloride) | 0.6 | 18.2 | 3.0 |
| Cream of tartar (potassium hydrogen tartrate) | 0.3 | 9.1 | 3.0 |

*Note:* Teaspoon measurements are approximations; when possible, use weight measurements for greater accuracy.

Table 1. Amounts of mordanting chemicals appropriate for treating 1 pound of wool, expressed in dry-weight quantities

Skeins can be keyed to indicate which mordant they are being treated with by tying knots in one end of the yarn. For example:

Alum = one knot
Copper = two knots
Iron = three knots
Tin = four knots

## Premordanting

Weigh the clean, dry yarns that are to be treated with a mordant. Record the weight. Submerge the yarns in room-temperature distilled or deionized water for thirty to forty-five minutes. Gently squeeze to push water into the yarns and remove air.

Using table 1, calculate and measure the appropriate amount of mordant to be used based upon the weight of the wool to be treated. Pour a few cups of distilled or deionized water into an enamel kettle. Add the correct amount of mordant and stir to dissolve. Add water to equal 4 gallons of water for every pound of wool.

Squeeze excess water from the soaking yarns and submerge the yarns into the mordant solution. Use rubber gloves and a heat-resistant spoon to prevent the mordant from coming in contact with your skin. Retain the soaking water for later use.

Place the mordant kettle on a stove and gradually raise the temperature to the point of a full steam or very low simmer. Be careful not to overheat. Open doors and windows, cover the kettle with a lid, and turn on an overhead vent or fan if possible. Do not stand directly over the kettle or inhale the fumes.

Steam or slightly simmer the yarn in the mordant solution for one to two hours. The colors reported in this book were achieved with two-hour mordanting; however, some dyers report heating the mordant solution for only one hour. Monitor the solution to prevent it from boiling. Occasionally swirl and turn the yarn over using a spoon, but avoid vigorous action. Add distilled or deionized water, beginning with the previously saved soaking water, to maintain at least 4 gallons of liquid per pound of wool.

Remove the kettle from the heat. If a hot dyebath is waiting, squeeze the excess mordant liquid from the yarns and transfer them to the dye. Otherwise leave the yarns in the mordant solution, allow them to cool gradually to room temperature, and continue with the premordant process. Some dyers believe that mordants are most effective if allowed to cure for a few days and therefore do not recommend immediately transferring yarns to a dyebath or rinsing with water. The dye results reported in this book were obtained with yarns that were mordanted and dried before dyeing.

If yarns are not to be dyed immediately, remove them from the cooled mordant solution. Gently squeeze out excess liquid and hang the skeins away from direct light, allowing them to drip-dry, or spread the yarns and dry flat. Another option is to refrigerate the damp skeins in a sealed container for later rinsing.

After twenty-four to forty-eight hours, submerge the mordanted yarns in three or four successive rinses of distilled or deionized water to remove excess mordant. If the yarns have been allowed to dry, begin with a thirty-minute soak to remoisten the fibers, and then proceed with the successive rinses. Allow the mordanted and rinsed yarns to dry (hanging or flat) away from light. Store the dried yarns in sealed containers, avoiding exposure to light in order to retain the full effectiveness of the mordant.

## Alternative Methods

As previously mentioned, a mordant may also be stirred into a dyebath before adding yarns, or dyed yarns may be submerged in a mordant solution after dyeing. Both are less effective than premordanting, however.

For either alternative, calculate the amount of mordant needed on the basis of the dry weight of the wool to be treated, using the figures provided in table 1. Be sure that the mordant is thoroughly dissolved before the wool is added. To avoid heat shock, the textile should be the same temperature as the dyebath or postmordant solution before being added to the liquid.

## Dyeing with Heat

Before dyeing, collect a sufficient amount of dyestuff for the quantity of wool to be colored. Most dyers recommend beginning with a one-to-one ratio of plant material to fiber, 1 ounce of dyestuff to 1 ounce of wool. Produce a test sample using this ratio, and if the color appears weak, increase the proportion of dyestuff.

As noted in chapter 1, natural dyeing is a process whereby a dyestuff is physically broken down, thereby releasing internal chromophores into water and producing a dyebath. Heating is one way to accomplish this breakdown. The following steps were used to achieve all of the heat-processed dye results reported in chapters 5 through 11.

Twenty-four hours before dyeing (or longer if using chunks of wood), begin preparing the dye. Wearing rubber gloves, divide the dye plant into its various components: flowers, fruit, leaves, and so forth. Treat each component as a distinct and separate dyestuff. Tear, chop, or cut the dye material into small pieces. Place the individual broken dyestuffs into separate enamel dyepots. These pans should be large enough to eventually hold 4 gallons of water for every pound of wool to be colored. Add enough distilled or deionized water to just cover the dyestuff. Set the pans aside overnight, allowing the water to soften the dye materials.

About an hour before dyeing (the following day), place the mordanted wool into a second enamel pan and cover it with distilled or deionized water. Squeeze the yarns to push water into the material and release trapped air. Wet fibers have much greater affinity for dye than do yarns added to the dyebath in a dry condition.

To prepare the dyebath, put the enamel pan containing a dyestuff and water onto the stove. Add water as needed to allow the pieces of plant material to move freely in the liquid. Slowly heat the dye material to a simmer or low boil. Maintain the simmer for a length of time determined by the tenderness of the plant part being used (see table 2). Add water if needed, making sure the dyestuff is always able to move freely in the liquid. Open doors and windows, cover the pan with a lid, or turn on an overhead vent or fan to reduce steam and fumes. Do not stand over the dyepot or inhale the vapors.

About fifteen minutes before dyeing (that is, fifteen minutes before the end of the timed simmer of the dyestuff), put the pan of wool and soaking water on the stove. Slowly heat the wool and liquid to steaming.

After the dyestuff has been simmered for the appropriate length of time, add distilled or deionized water to the dye in proportion to the dry weight of the wool to be colored. Each pound of wool should have approximately 4 gallons of dye liquid. If the additional water cools the dyebath, reheat the dyebath to achieve and maintain a very low simmer. Some dyers prefer to strain vegetable matter from the dye liquid at this point, before adding the

Flowers of *Castilleja indivisa* are torn, the pieces falling into an enamel pan for overnight soaking.

Before the yarns are added, the dye material is simmered to release the chromophores.

| PLANT ORGAN | LENGTH OF SIMMER (MINUTES) |
|---|---|
| Flowers | 20 |
| Fruits | 20 |
| Leaves | |
| Tender and soft | 15 |
| Semi-tough | 25 |
| Tough | 35 |
| Roots | |
| Herbaceous | 45 |
| Woody | 120 |
| Herbaceous stems | |
| Tender and soft | 15 |
| Semi-tough | 30 |
| Tough | 45 |
| Woody stems and bark | 120 |

Table 2. Recommended heating times for plant dyestuffs, prior to the addition of the textile

textile. Doing so prevents the yarns from tangling on and adhering to pieces of dyestuff, but in the case of weak dyes this may result in less color in the dyebath.

When the dyebath is at a very low simmer, lift the heated wool from its pan (make sure to wear rubber gloves) and squeeze out excess water. Place the wool in the dyebath. To achieve full color potential, keep the dyebath steaming for at least twenty-five minutes, but do not boil. For lighter hues, monitor the absorption of color and reduce the dyeing time as needed. Remember that wet wool will appear slightly darker than it will be when dry. Longer dyeing times (up to an hour) may produce greater colorfastness.

After simmering the textile on the stove for twenty-five minutes to an hour (or less time for lighter hues), remove the dyepot from the heat. Allow the dyebath, with yarns still submerged, to cool gradually to room temperature. When the dyebath has cooled, remove and rinse the dyed yarns in successive baths of distilled or deionized water until the rinse water remains clear. Some dyers gently wash the textile at this point, using a mild detergent. Doing so may affect the color but also may reduce the likelihood of further fading if the finished material is to be laundered in the future.

Hang or otherwise position the dyed yarn skeins away from direct light and allow them to dry. Strain all vegetable matter from the dyebath if this was not done at an earlier stage. Discard the dye liquid down a bathroom sink or toilet, and throw the vegetable matter into the garbage. Label each dried skein with the name of the dyestuff, the mordant, and any other information that may be needed for future reference. Store the dyed yarns away from light and high humidity to reduce the likelihood of fading.

If a particularly pleasing hue is obtained, carefully record all aspects of the dyeing process (collection date, plant component, proportion of plant material to wool, mordant, and so on) in a journal for future use. Remember, however, that some color variability is common when using natural dyes: don't expect exact duplication.

After the dyebath has cooled, the yarns are removed and rinsed in distilled water.

## Dyeing with Decomposition

One of the most primitive methods of dyeing involves decomposition. Plant matter is placed in a tepid liquid, usually water, and allowed to decay, thereby releasing chromophores. Historically, additives such as urine, sugar, yeast, or bran were sometimes stirred into the decomposing material to enhance either spoilage or fermentation. A modern variation of this decomposition process is solar dyeing, in which the water and plant material are placed into a glass jar and set in direct sunlight. The sun's heat enhances decomposition, but the light rays may also influence the color produced. When plant matter is decomposed in a wooden container, tannic acid from the wood is released into the solution, which darkens the color and also serves as a mordant. The following procedure was used for all of the decomposition results reported in chapters 5 through 11.

Wearing rubber gloves, separate a dye plant into its various components, such as flowers, fruits, and leaves. Treat each plant part as a distinct and separate dyestuff. Tear, chop, or cut the dyestuff into small pieces and place them in a glass jar or plastic tub for which a tight-fitting lid is available.

Measure and add sufficient distilled or deionized water to allow the dyestuff to move freely in the liquid. Record the amount of water used. Add a quarter teaspoon of granulated sugar to the container for each cup of water that was poured over the dye material. If using a jar with a metal lid, place a small plastic bag over the jar mouth to prevent the dye liquid from coming into contact with the metal. Securely fasten the lid. Shake the container to mix the water, sugar, and plant material. Place tape on the outside of the container and on it record the plant name, plant component, and date.

Place the container in a box or on a dark shelf. Periodically open the container (approximately once each week)

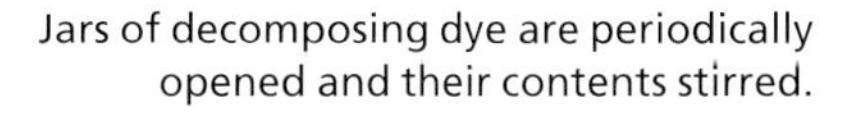

Jars of decomposing dye are periodically opened and their contents stirred.

and stir the contents. When the liquid contains leached color, and this color does not strengthen with additional time, the dye is ready to be used.

Thirty to forty-five minutes before dyeing, submerge mordanted yarns in distilled or deionized water. Gently squeeze the yarns to remove trapped air and push water into the material, remembering that wet fibers have much greater affinity for dye than do yarns added to the dyebath in a dry condition. Strain the dye liquid into a clean container and discard the decaying plant matter in the garbage. (It will be very odiferous, so make sure there is good ventilation!) Add distilled or deionized water as needed to create a dyebath in which the textile can move freely. Squeeze excess water from the soaking wool, and gently stir the damp yarns into the dyebath. Securely seal the dye container, again applying a small plastic bag if the container lid is metal. Transfer the tape label from the original container, and add the date upon which the yarns were placed into the dye solution.

Once or twice on each succeeding day, either stir or gently shake the container contents to reposition the wool in the dyebath. Monitor the color of the yarns. For maximum color absorption, leave the yarns in the dye for up to seven days (fewer days if the desired color has been achieved).

Rinse the dyed yarns in successive baths of distilled or deionized water until the rinse water is clear. Remove the yarn from the last rinse and add a small amount of mild detergent. Mix the detergent into the water and resubmerge the yarns. Gently squeeze the textile to force the soapy water into the yarns. Although this wash may affect the color of the yarns, it is necessary to remove the foul odor associated with decomposed dye solutions. Rinse the washed yarns in successive baths of distilled or deionized water until all evidence of detergent is removed.

Dry the yarn skeins away from direct light. Label each dry skein with the name of the dyestuff, the mordant, and any other information that may be needed

Excess water is squeezed from the soaking wool, and the damp yarns are gently stirred into the dyebath.

for future reference. Store the yarn away from light and high humidity to reduce the likelihood of fading.

## Fastness

Fastness refers to the degree to which a dyed textile resists fading. The more durable the color, the greater the colorfastness. Dyes are most prone to fading when exposed to light and moisture. Information concerning a dye's fastness facilitates selection of the best use for a textile colored with that dye. Dyes that fade when wet should not be used for textile products that will be laundered; dry cleaning can usually be employed to clean material dyed with these delicate colors. Similarly, dyes that are sensitive to light should not be used to color fabrics that will be displayed near sunlight or fluorescent lights. Pretesting dyed yarns for fastness can eliminate unpleasant surprises.

### Testing for Lightfastness

Cut several 14-inch strands of yarn that have been colored with the same dyestuff, mordant, and dye process. Position these yarns lengthwise on a piece of cardboard, preferably white, which has been cut approximately 16 inches long by 4 inches wide. Position the yarns so that they are about an eighth of an inch apart, and tape the yarn ends to hold them in place. Cut another piece of cardboard 13 inches long and 4 inches wide. Position this second piece of cardboard on top of the first so that the top cardboard covers all but 3 inches of the length of the yarns, with about 11 inches covered. Fasten the two pieces of cardboard together in this position with sturdy clips or pieces of tape. Place the device in direct sunlight for ten hours. It may take several days to reach this total sun exposure time.

After exposure, unclip the top cardboard and reposition it, exposing about 5 inches of yarn and leaving about 9 inches of yarn covered. Refasten the two boards and place the device in direct sunlight for another ten hours. Repeat this sequence three more times, moving the cardboard 2 inches at a time until all but the last 3 inches are exposed to direct sunlight.

Remove the top cardboard and analyze the results. Looking at all 14 inches of yarn, you will see a record of the lightfastness of the dye being tested. If the top and bottom ends of the yarn are similar in color, the yarn has good lightfastness. However, if these vary, the dye is fugitive. Also examine the surface of the cardboard that was touching the yarns. If the cardboard has picked up color from the yarns, this indicates that the dye tends to migrate, or transfer, to adjacent surfaces. Care will be needed in storing

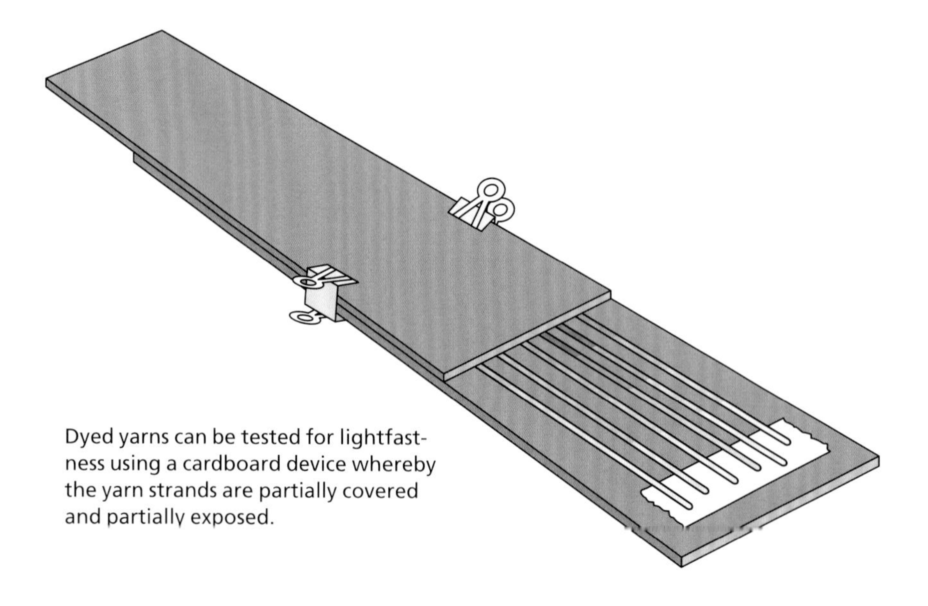

Dyed yarns can be tested for lightfastness using a cardboard device whereby the yarn strands are partially covered and partially exposed.

and using objects made from yarns containing this kind of dye, because they may discolor surrounding textiles.

## Testing for Washfastness

Soak a piece of dyed yarn in warm water in a clear glass jar. Note any changes in the color of the water. Remove the yarn and mix a little detergent into the water. Return the yarn to the water and stir gently. Let the yarn soak in the soapy solution for about fifteen minutes, and again stir it gently. Observe the color of the wash water to detect the loss of dye from the yarn. Transfer the yarn to a jar of clean rinse water. Agitate the sample to remove the detergent and watch for evidence of bleeding color.

Transfer the rinsed yarn to a white paper towel. Fold the towel over the sample and press out excess water. Check the towel for the presence of color from the yarn. Finally, compare the washed sample with an unwashed sample colored with the same dye solution, mordant, and dye process. If these two yarns still have similar color, the dye is fairly washfast. However, if the two samples are obviously different, the dye is fugitive and best suited for uses that will not require washing.

# 5 PURPLE AND RED

## Nature's Rarest Dye Colors

Our search for natural dyes resulted in various purple colors, shown here with Munsell notations. Following these color chips is a list of the plants that were used to obtain these colors, including mordant and processing conditions.

5R 4/1
**DARK GRAYED RED·VIOLET**

5RP 7/1 7.5R 7/2
**GRAYED RED·VIOLET**

7.5P 7/2
**GRAYED VIOLET**

2.5YR 8/2

5YR 8/1

7.5R 8/2

**LIGHT GRAYED RED·VIOLET**

7.5RP 7/4

**RED·VIOLET**

| DYE SOURCE | DYE RESULT | MORDANT | PROCESSING |
|---|---|---|---|
| ***Ceanothus americanus*** NEW JERSEY TEA | | | |
| rhizomes and roots | dark grayed red·violet | iron | heat |
| ***Gaillardia pulchella*** INDIAN BLANKET | | | |
| ray florets | light grayed red·violet | none | decomposition |
| ***Robinia pseudoacacia*** BLACK LOCUST | | | |
| leaves | red·violet | tin | decomposition |
| ***Rudbeckia hirta*** BLACK-EYED SUSAN | | | |
| disk florets | grayed red·violet | alum | decomposition |
| ***Sassafras albidum*** SASSAFRAS | | | |
| roots | light grayed red·violet | alum | heat |
| ***Vitis aestivalis*** PIGEON GRAPE | | | |
| berries | light grayed red·violet | alum | heat |
| | light grayed red·violet | copper | heat |
| | grayed red·violet | iron | heat |
| | grayed violet | tin | decomposition |

Our search for natural dyes resulted in various red colors, shown here with Munsell notations. Following these color chips is a list of the plants that were used to obtain these colors, including mordant and processing conditions.

7.5R 3/6

**BURGUNDY·RED**

10R 6/6

**DARK SALMON·ROSE**

7.5YR 8/4

**PEACH**

10R 8/4

**PEACH·PINK**

10YR 9/2
**PALE PEACH**

5YR 8/4
**PINK·PEACH**

10R 7/6
**SALMON·PINK**

| DYE SOURCE | DYE RESULT | MORDANT | PROCESSING |
|---|---|---|---|
| ***Acer saccharinum*** SILVER MAPLE | | | |
| bark | salmon·pink | alum | decomposition |
| | peach·pink | tin | decomposition |
| ***Asimina triloba*** PAWPAW | | | |
| leaves | pale peach | alum | decomposition |
| ***Castanea ozarkensis*** OZARK CHINKAPIN | | | |
| branches | dark salmon·rose | tin | decomposition |
| | peach | tin | heat |
| ***Castilleja indivisa*** INDIAN PAINTBRUSH | | | |
| bract apices | pale peach | none | heat |
| ***Ceanothus americanus*** NEW JERSEY TEA | | | |
| stems | peach | alum | decomposition |
| ***Cephalanthus occidentalis*** BUTTONBUSH | | | |
| branches | peach | tin | decomposition |
| | pale peach | none | heat |
| ***Cercis canadensis*** REDBUD | | | |
| roots | peach | alum | decomposition |
| ***Dracopis amplexicaulis*** CLASPING-LEAVED CONEFLOWER | | | |
| stems | pale peach | none | decomposition |
| ***Ipomopsis rubra*** STANDING CYPRESS | | | |
| petals | pale peach | none | heat |
| ***Juniperus virginiana*** EASTERN RED CEDAR | | | |
| bark | peach | alum | heat |
| | peach | none | heat |
| ***Liquidambar styraciflua*** SWEET GUM | | | |
| bark | burgundy·red | tin | decomposition |
| ***Oenothera heterophylla*** SAND EVENING-PRIMROSE | | | |
| roots | salmon·pink | tin | decomposition |
| ***Polytaenia nuttallii*** PRAIRIE PARSLEY | | | |
| flowers and pedicels | pale peach | none | decomposition |
| ***Rhus glabra*** SMOOTH SUMAC | | | |
| fruits | pink·peach | alum | decomposition |

| DYE SOURCE | DYE RESULT | MORDANT | PROCESSING |
|---|---|---|---|
| ***Salix caroliniana*** CAROLINA WILLOW | | | |
| branches | peach | alum | heat |
| | pale peach | alum | decomposition |
| | pale peach | tin | decomposition |
| | pale peach | none | heat |
| | pale peach | none | decomposition |
| ***Salix nigra*** BLACK WILLOW | | | |
| bark | peach | alum | heat |
| | pale peach | none | heat |
| ***Stillingia sylvatica*** QUEEN'S DELIGHT | | | |
| roots | pale peach | none | decomposition |
| ***Ulmus rubra*** SLIPPERY ELM | | | |
| leaves | peach | tin | decomposition |
| ***Vaccinium arboreum*** FARKLEBERRY | | | |
| rootstocks | pale peach | none | heat |
| ***Viburnum rufidulum*** RUSTY BLACK HAW | | | |
| bark | peach | alum | decomposition |
| | peach | none | heat |

# Nature's Own Color

Our search for natural dyes resulted in various green colors, shown here with Munsell notations. Following these color chips is a list of the plants that were used to obtain these colors, including mordant and processing conditions.

7.5Y 6/10

10Y 7/8

**ACID YELLOW·GREEN**

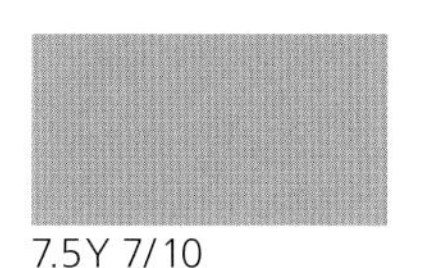

7.5Y 7/10

7.5Y 7/12

10Y 7/12

**BRIGHT ACID YELLOW·GREEN**

5Y 5/8

**BRIGHT MEDIUM OLIVE**

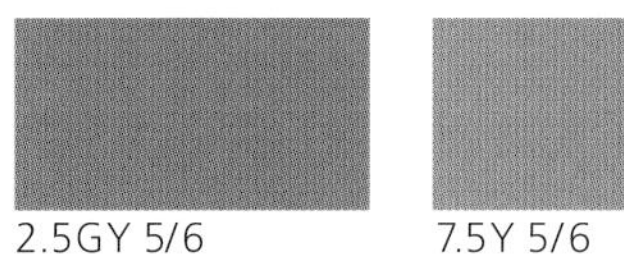

2.5GY 5/6 7.5Y 5/6

**BRIGHT OLIVE**

2.5GY 6/8

**BRIGHT YELLOW·GREEN**

 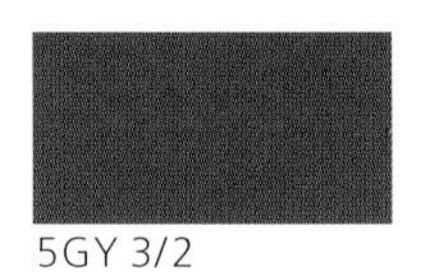 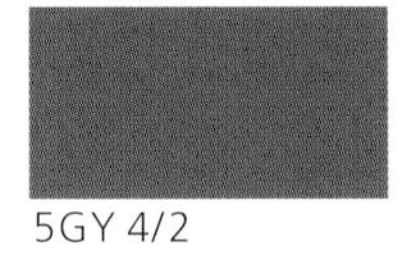 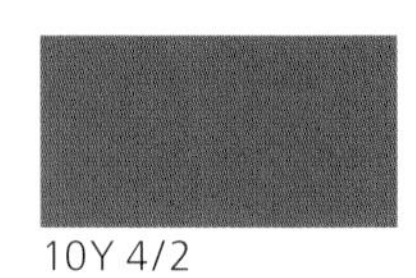

2.5GY 3/2 5GY 3/2 5GY 4/2 10Y 4/2

**DARK GRAY·GREEN**

  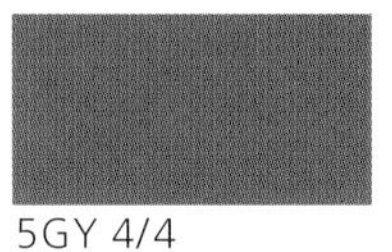

2.5G 4/4 5GY 4/4

**DARK GREEN**

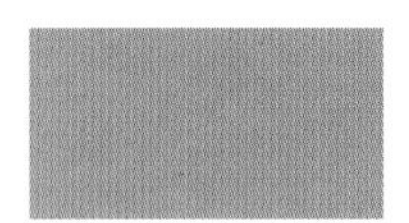

5Y 5/4

**DARK KHAKI·GREEN**

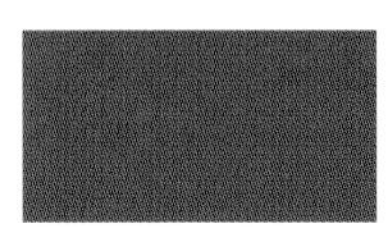 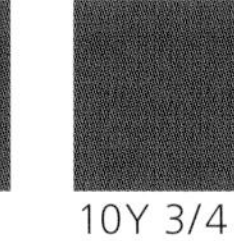

7.5Y 3/4 10Y 3/4

**DARK OLIVE**

5GY 6/6

**DULL LIME·GREEN**

7.5GY 7/4

**DULL MEDIUM GREEN**

7.5GY 3/2

**GRAYED FOREST·GREEN**

7.5Y 7/8

**LIGHT ACID YELLOW·GREEN**

7.5Y 6/6 7.5Y 7/6

**LIGHT BRIGHT OLIVE**

2.5GY 9/6

**LIGHT BRIGHT YELLOW·GREEN**

5GY 7/4

**LIGHT DULL GREEN**

2.5GY 8/4

**LIGHT DULL YELLOW·GREEN**

2.5GY 7/4 5GY 8/2 7.5GY 8/2

**LIGHT GRAY·GREEN**

10GY 9/2

**LIGHT GREEN**

5Y 6/6 7.5Y 6/4 7.5Y 7/4 10Y 6/6

**LIGHT OLIVE**

10Y 7/4 10Y 7/6

**LIGHT OLIVE, continued**

2.5GY 7/6 2.5GY 8/6

**LIGHT YELLOW·GREEN**

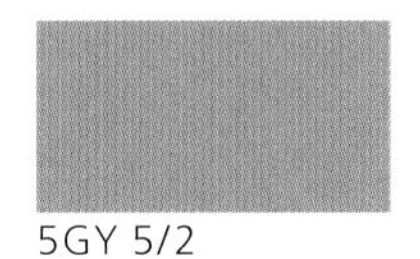

5GY 5/2

**MEDIUM DULL GREEN**

2.5GY 5/2 2.5GY 6/4 10Y 5/2

**MEDIUM GRAY·GREEN**

5Y 6/4

**MEDIUM KHAKI·GREEN**

2.5GY 5/4 5Y 4/6 10Y 5/4

**MEDIUM OLIVE**

2.5GY 4/4 5Y 5/6 7.5Y 4/4 7.5Y 4/6

**OLIVE**

7.5Y 5/4

**OLIVE, continued**

2.5GY 9/4 5GY 9/4

**PALE YELLOW·GREEN**

7.5G 6/2

**TEAL·GREEN**

10Y 3/2

**VERY DARK OLIVE**

| DYE SOURCE | DYE RESULT | MORDANT | PROCESSING |
|---|---|---|---|
| ***Acer negundo*** BOXELDER | | | |
| bark | medium khaki·green | iron | heat |
| ***Achillea millefolium*** YARROW | | | |
| leaves | light olive | iron | decomposition |
| ***Ageratina altissima*** WHITE SNAKEROOT | | | |
| heads | medium olive | iron | heat |
| leaves | dark khaki·green | iron | heat |
| roots | light olive | iron | heat |
| stems | light olive | iron | heat |
| ***Alnus maritima*** SEASIDE ALDER | | | |
| leaves | light olive | iron | heat |
| ***Ambrosia psilostachya*** WESTERN RAGWEED | | | |
| leaves and heads | bright acid yellow·green | copper | heat |
| | light olive | iron | decomposition |
| ***Ambrosia trifida*** GIANT RAGWEED | | | |
| leaves | light olive | copper | heat |
| roots | light olive | iron | heat |
| stems | light dull yellow·green | alum | heat |
| ***Amorpha canescens*** LEADPLANT | | | |
| flowers | olive | iron | heat |
| | medium khaki·green | iron | decomposition |
| ***Apocynum cannabinum*** INDIAN HEMP | | | |
| leaves | bright olive | iron | decomposition |
| ***Argemone polyanthemos*** PRICKLY POPPY | | | |
| leaves | medium khaki·green | copper | heat |
| | dark khaki·green | iron | heat |
| stems | dark khaki·green | copper | heat |
| ***Asclepias tuberosa*** BUTTERFLY MILKWEED | | | |
| flowers | light bright olive | iron | heat |
| leaves | medium olive | iron | heat |
| ***Asclepias viridis*** ANTELOPE-HORN MILKWEED | | | |
| leaves | light bright olive | copper | heat |
| ***Asimina triloba*** PAWPAW | | | |
| leaves | dark khaki·green | iron | heat |
| ***Baccharis salicina*** WILLOW BACCHARIS | | | |
| flowers | medium khaki·green | copper | decomposition |
| leaves | olive | iron | heat |
| | dull medium green | iron | decomposition |
| stems | medium khaki·green | iron | heat |
| ***Baptisia australis*** BLUE WILD-INDIGO | | | |
| leaves and upper stems | light acid yellow·green | alum | heat |
| | olive | copper | heat |
| | light acid yellow·green | tin | heat |
| | light olive | none | decomposition |
| petals | medium khaki·green | iron | heat |
| stems (basal) | medium khaki·green | iron | heat |
| ***Baptisia bracteata*** PLAINS WILD-INDIGO | | | |
| leaves | light olive | copper | heat |

| DYE SOURCE | DYE RESULT | MORDANT | PROCESSING |
|---|---|---|---|
| ***Bidens bipinnata*** SPANISH NEEDLES | | | |
| heads | medium khaki·green | copper | decomposition |
| leaves | bright acid yellow·green | none | heat |
| roots | light olive | copper | heat |
| | dark gray·green | iron | heat |
| stems | light bright olive | copper | heat |
| | olive | iron | heat |
| ***Brickellia eupatorioides*** FALSE BONESET | | | |
| heads | light acid yellow·green | copper | heat |
| | olive | iron | heat |
| leaves | light acid yellow·green | copper | heat |
| | olive | iron | heat |
| roots | light olive | iron | heat |
| ***Buchnera americana*** BLUEHEARTS | | | |
| flowers | light olive | tin | heat |
| leaves | acid yellow·green | alum | heat |
| | light olive | copper | heat |
| | acid yellow·green | tin | heat |
| ***Bumelia lanuginosa*** CHITTAMWOOD | | | |
| fruits | dull medium green | copper | heat |
| | teal·green | iron | heat |
| | very dark olive | tin | heat |
| ***Callicarpa americana*** AMERICAN BEAUTYBERRY | | | |
| branchlets | light olive | iron | heat |
| ***Calylophus serrulatus*** HALFSHRUB SUNDROP | | | |
| leaves | olive | copper | heat |
| | olive | iron | decomposition |
| petals | medium khaki·green | iron | heat |
| stems | medium khaki·green | iron | heat |
| ***Campsis radicans*** TRUMPET CREEPER | | | |
| flowers | light olive | tin | heat |
| fruits | light olive | iron | heat |
| leaves | olive | iron | heat |
| sepals | light bright olive | iron | heat |
| ***Castilleja indivisa*** INDIAN PAINTBRUSH | | | |
| leaves and stems | pale yellow·green | alum | heat |
| | light olive | copper | heat |
| | olive | iron | heat |
| | dull lime·green | tin | heat |
| ***Ceanothus americanus*** NEW JERSEY TEA | | | |
| flowers | dark khaki·green | iron | heat |
| leaves | medium khaki·green | iron | heat |
| ***Centaurea americana*** BASKET FLOWER | | | |
| leaves and stems | bright acid yellow-green | alum | heat |
| | acid yellow-green | copper | heat |
| | olive | iron | heat |
| ***Cephalanthus occidentalis*** BUTTONBUSH | | | |
| branches | light olive | copper | decomposition |

| DYE SOURCE | DYE RESULT | MORDANT | PROCESSING |
|---|---|---|---|
| ***Cercis canadensis*** REDBUD | | | |
| flowers | light olive | iron | heat |
| ***Chrysopsis pilosa*** SOFT GOLDEN-ASTER | | | |
| leaves | bright olive | copper | heat |
| petals | bright acid yellow-green | alum | heat |
| | bright olive | copper | heat |
| roots | light gray-green | copper | heat |
| | medium gray-green | iron | heat |
| ***Cirsium undulatum*** WAVY-LEAF THISTLE | | | |
| leaves | dark khaki-green | iron | heat |
| ***Coreopsis tinctoria*** PLAINS TICKSEED | | | |
| leaves | dark khaki-green | iron | decomposition |
| ray florets | bright medium olive | copper | heat |
| ***Cornus drummondii*** ROUGH-LEAF DOGWOOD | | | |
| branchlets | medium khaki-green | copper | heat |
| | very dark olive | iron | heat |
| fruits | light olive | copper | heat |
| leaves | light bright olive | copper | heat |
| roots | medium olive | copper | heat |
| ***Cornus florida*** FLOWERING DOGWOOD | | | |
| fruits | light olive | iron | heat |
| ***Cucurbita foetidissima*** BUFFALO GOURD | | | |
| leaves | medium khaki-green | copper | decomposition |
| ***Cycloloma atriplicifolium*** TUMBLE RINGWING | | | |
| leaves and spikes | olive | iron | heat |
| | light olive | iron | decomposition |
| stems | olive | iron | heat |
| ***Cyperus squarrosus*** BEARDED FLATSEDGE | | | |
| spikes | medium khaki-green | iron | heat |
| ***Dalea candida*** WHITE PRAIRIE-CLOVER | | | |
| leaves | light olive | copper | decomposition |
| | medium olive | iron | heat |
| | light olive | tin | decomposition |
| spikes | light bright olive | copper | decomposition |
| | olive | iron | heat |
| | olive | iron | decomposition |
| ***Dalea lanata*** WOOLLY DALEA | | | |
| flowers | olive | iron | heat |
| | medium khaki-green | iron | decomposition |
| leaves | light bright olive | copper | decomposition |
| | light olive | iron | decomposition |
| ***Dalea purpurea*** PURPLE PRAIRIE-CLOVER | | | |
| leaves and stems | olive | copper | decomposition |
| | light olive | iron | heat |
| | dark khaki-green | iron | decomposition |
| ***Dalea villosa*** SILKY PRAIRIE-CLOVER | | | |
| flowers | medium khaki-green | iron | heat |
| leaves | medium khaki-green | iron | decomposition |

| DYE SOURCE | DYE RESULT | MORDANT | PROCESSING |
|---|---|---|---|
| ***Desmanthus illinoensis*** ILLINOIS BUNDLEFLOWER | | | |
| fruits | dark khaki-green | copper | heat |
| leaves | dark khaki-green | copper | heat |
| stems | medium khaki-green | copper | heat |
| ***Dracopis amplexicaulis*** CLASPING-LEAVED CONEFLOWER | | | |
| leaves | dark olive | iron | heat |
| ray florets | light olive | iron | heat |
| roots | light olive | iron | heat |
| stems | olive | iron | heat |
| ***Echinacea angustifolia*** PURPLE PRAIRIE CONEFLOWER | | | |
| disk florets | very dark olive | iron | heat |
| leaves | light olive | iron | decomposition |
| ray florets | dark khaki-green | iron | heat |
| ***Echinocereus reichenbachii*** LACE HEDGEHOG-CACTUS | | | |
| flowers | medium khaki-green | iron | heat |
| ***Elephantopus carolinianus*** LEAFY ELEPHANT'S FOOT | | | |
| leaves | light olive | iron | decomposition |
| roots | light olive | copper | heat |
| | medium gray-green | iron | heat |
| | medium khaki-green | iron | decomposition |
| stems | medium khaki-green | copper | decomposition |
| | medium khaki-green | iron | decomposition |
| ***Euphorbia maculata*** SPOTTED SPURGE | | | |
| leaves and stems | medium khaki-green | copper | heat |
| ***Euphorbia marginata*** SNOW-ON-THE-MOUNTAIN | | | |
| bracts | light bright yellow-green | copper | heat |
| cyathia | light olive | iron | decomposition |
| leaves | light bright olive | iron | heat |
| roots | medium olive | copper | heat |
| | medium gray-green | iron | heat |
| stems | light olive | copper | heat |
| | medium gray-green | iron | heat |
| | bright acid yellow-green | tin | heat |
| ***Fraxinus pennsylvanica*** GREEN ASH | | | |
| bark | light olive | copper | heat |
| | olive | iron | heat |
| leaves | light olive | copper | heat |
| ***Fraxinus quadrangulata*** BLUE ASH | | | |
| bark | olive | copper | heat |
| | dark olive | iron | heat |
| leaves | dark khaki-green | iron | heat |
| | light olive | iron | decomposition |
| ***Gaillardia aestivalis*** PRAIRIE GAILLARDIA | | | |
| disk florets | acid yellow-green | alum | heat |
| | light bright olive | copper | heat |
| | dark olive | iron | heat |
| | olive | iron | decomposition |
| | dark green | tin | heat |

| DYE SOURCE | DYE RESULT | MORDANT | PROCESSING |
|---|---|---|---|
| ***Gaillardia aestivalis,*** continued | | | |
| leaves | bright acid yellow-green | alum | heat |
| | bright medium olive | copper | heat |
| | light bright olive | iron | decomposition |
| ray florets | medium khaki-green | iron | heat |
| | light yellow-green | tin | heat |
| ***Gaillardia pulchella*** INDIAN BLANKET | | | |
| ray florets | light olive | tin | heat |
| ***Glandularia canadensis*** ROSE VERVAIN | | | |
| flowers | dull lime-green | tin | heat |
| leaves and stems | light olive | iron | heat |
| | bright yellow-green | tin | heat |
| | light dull yellow-green | none | heat |
| ***Gymnocladus dioicus*** KENTUCKY COFFEE-TREE | | | |
| leaflets | light olive | iron | heat |
| ***Helenium amarum*** BITTER SNEEZEWEED | | | |
| heads | light olive | iron | heat |
| stems | light olive | iron | heat |
| ***Helianthus annuus*** ANNUAL SUNFLOWER | | | |
| disk florets | light olive | copper | heat |
| | dark gray-green | iron | heat |
| leaves | light acid yellow-green | alum | heat |
| | olive | copper | heat |
| ray florets | light olive | iron | heat |
| ***Helianthus maximiliani*** MAXIMILIAN'S SUNFLOWER | | | |
| disk florets and phyllaries | olive | iron | heat |
| ray florets | light bright olive | iron | heat |
| ***Helianthus mollis*** ASHY SUNFLOWER | | | |
| disk florets | light bright olive | copper | heat |
| leaves | olive | copper | heat |
| ray florets | light olive | copper | heat |
| | medium gray-green | iron | heat |
| ***Helianthus tuberosus*** JERUSALEM ARTICHOKE | | | |
| disk and ray florets | light acid yellow-green | copper | heat |
| | olive | iron | heat |
| leaves | acid yellow-green | copper | heat |
| | olive | iron | heat |
| | medium khaki-green | iron | decomposition |
| roots | light olive | copper | heat |
| | light olive | iron | heat |
| stems | light olive | iron | heat |
| ***Hydrangea arborescens*** WILD HYDRANGEA | | | |
| flowers | medium khaki-green | iron | heat |
| leaves | medium khaki-green | iron | heat |
| ***Ipomoea leptophylla*** BUSH MORNING-GLORY | | | |
| leaves | dark khaki-green | iron | heat |
| ***Ipomopsis rubra*** STANDING CYPRESS | | | |
| leaves | dark khaki-green | iron | heat |

| DYE SOURCE | DYE RESULT | MORDANT | PROCESSING |
|---|---|---|---|
| ***Juniperus virginiana*** EASTERN RED CEDAR | | | |
| leaves and stems | light olive | iron | heat |
| ***Liatris punctata*** DOTTED GAYFEATHER | | | |
| leaves and stems | medium khaki-green | iron | decomposition |
| ***Liatris squarrosa*** SCALY GAYFEATHER | | | |
| flowers | light olive | iron | heat |
| | medium dull green | tin | heat |
| leaves | light bright olive | copper | heat |
| | olive | iron | heat |
| ***Linum sulcatum*** GROOVED FLAX | | | |
| leaves, stems, and petals | light bright olive | iron | decomposition |
| ***Liquidambar styraciflua*** SWEET GUM | | | |
| leaves | medium khaki-green | copper | heat |
| roots | light olive | copper | heat |
| ***Maclura pomifera*** OSAGE ORANGE | | | |
| bark | dark khaki-green | iron | heat |
| ***Mentzelia nuda*** BRACTLESS BLAZING-STAR | | | |
| flowers | light bright olive | copper | heat |
| ***Monarda fistulosa*** WILD BERGAMOT | | | |
| leaves | light acid yellow-green | alum | heat |
| | light bright olive | copper | heat |
| | dark olive | iron | heat |
| | light olive | iron | decomposition |
| | bright acid yellow-green | tin | heat |
| spikes | light bright olive | copper | heat |
| | olive | iron | heat |
| | bright acid yellow-green | tin | heat |
| ***Morus rubra*** RED MULBERRY | | | |
| bark green | light olive | iron | heat |
| branchlets | olive | iron | heat |
| ***Neptunia lutea*** YELLOW NEPTUNE | | | |
| flowers | olive | iron | heat |
| leaves | light olive | copper | heat |
| | light bright olive | copper | decomposition |
| | olive | iron | decomposition |
| spikes | olive | iron | heat |
| ***Nyssa sylvatica*** BLACK GUM | | | |
| leaves | light olive | copper | heat |
| | olive | iron | decomposition |
| ***Oenothera heterophylla*** SAND EVENING-PRIMROSE | | | |
| leaves | light bright olive | copper | heat |
| ***Oenothera macrocarpa*** BIGFRUIT EVENING-PRIMROSE | | | |
| bracts and stems | medium khaki-green | iron | heat |
| fruits mature | medium khaki-green | iron | heat |
| fruits immature | dark gray-green | iron | heat |
| leaves | dark khaki-green | iron | decomposition |
| petals | light olive | iron | heat |

| DYE SOURCE | DYE RESULT | MORDANT | PROCESSING |
|---|---|---|---|
| ***Packera obovata*** ROUNDLEAF GROUNDSEL | | | |
| flowers | light bright olive | iron | heat |
| leaves and stems | olive | iron | heat |
| ***Parthenocissus quinquefolia*** VIRGINIA CREEPER | | | |
| fruits | light olive | copper | heat |
| | medium gray-green | iron | heat |
| ***Pediomelum cuspidatum*** TALL-BREAD SCURFPEA | | | |
| leaves | light olive | iron | heat |
| ***Phlox pilosa*** PRAIRIE PHLOX | | | |
| leaves | medium khaki-green | iron | heat |
| petals | light acid yellow-green | copper | heat |
| | dark khaki-green | iron | heat |
| ***Phyla lanceolata*** LANCELEAF FROG-FRUIT | | | |
| flowers | medium khaki-green | iron | heat |
| ***Physalis virginiana*** VIRGINIA GROUND-CHERRY | | | |
| leaves | light olive | iron | heat |
| ***Polygonum pensylvanicum*** PENNSYLVANIA SMARTWEED | | | |
| flowers and flower buds | medium olive | copper | heat |
| | medium khaki-green | iron | decomposition |
| leaves | light olive | alum | heat |
| | medium gray-green | copper | heat |
| | dark gray-green | iron | heat |
| stems | medium gray-green | iron | heat |
| ***Polygonum punctatum*** DOTTED SMARTWEED | | | |
| flowers and flower buds | light olive | copper | heat |
| | olive | iron | heat |
| leaves | light olive | copper | heat |
| | olive | iron | heat |
| | light olive | iron | decomposition |
| stems | medium olive | iron | heat |
| | light olive | iron | decomposition |
| ***Potentilla arguta*** TALL CINQUEFOIL | | | |
| leaves | light olive | copper | heat |
| stems | olive | copper | heat |
| ***Psoralidium tenuiflorum*** SLIM-FLOWER SCURFPEA | | | |
| flowers | acid yellow-green | tin | heat |
| leaves | medium khaki-green | copper | decomposition |
| ***Ptilimnium nuttallii*** MOCK BISHOP'S-WEED | | | |
| leaves and stems | light olive | copper | heat |
| ***Pyrrhopappus grandiflorus*** MORNING STAR | | | |
| leaves | dark olive | iron | heat |
| ***Ratibida columnifera*** MEXICAN HAT | | | |
| disk florets | light gray-green | alum | heat |
| | dark gray-green | iron | heat |
| ***Rhus copallinum*** WINGED SUMAC | | | |
| bark | dark khaki-green | copper | heat |
| flowers | dark khaki-green | copper | heat |
| leaves | light olive | iron | decomposition |

| DYE SOURCE | DYE RESULT | MORDANT | PROCESSING |
|---|---|---|---|
| ***Rhus glabra*** SMOOTH SUMAC | | | |
| leaves | dark khaki-green | copper | heat |
| ***Robinia hispida*** BRISTLY LOCUST | | | |
| leaves | light olive | copper | heat |
| | medium khaki-green | iron | decomposition |
| ***Robinia pseudoacacia*** BLACK LOCUST | | | |
| leaves | light olive | copper | heat |
| ***Rudbeckia grandiflora*** ROUGH CONEFLOWER | | | |
| ray florets | medium khaki-green | iron | heat |
| rhizomes | light olive | iron | heat |
| ***Rudbeckia hirta*** BLACK-EYED SUSAN | | | |
| disk florets | medium gray-green | copper | heat |
| | grayed forest-green | iron | heat |
| leaves | light olive | iron | decomposition |
| roots | light bright olive | iron | heat |
| ***Rumex altissimus*** PALE DOCK | | | |
| leaves | dark khaki-green | iron | heat |
| stems | bright olive | copper | heat |
| ***Sabatia campestris*** PRAIRIE ROSE GENTIAN | | | |
| petals | light gray-green | tin | heat |
| ***Salix caroliniana*** CAROLINA WILLOW | | | |
| leaves | medium khaki-green | iron | heat |
| | light olive | iron | decomposition |
| ***Salix exigua*** SANDBAR WILLOW | | | |
| leaves | light bright olive | copper | heat |
| | light olive | iron | heat |
| ***Salix nigra*** BLACK WILLOW | | | |
| leaves | olive | iron | heat |
| ***Sapindus drummondii*** SOAPBERRY | | | |
| flowers | olive | iron | heat |
| leaves | light bright olive | copper | heat |
| ***Sassafras albidum*** SASSAFRAS | | | |
| flowers | light olive | iron | decomposition |
| leaves | light bright olive | copper | heat |
| | olive | iron | heat |
| ***Silphium integrifolium*** PRAIRIE ROSINWEED | | | |
| disk and ray florets | light olive | iron | heat |
| roots | olive | iron | heat |
| ***Sisyrinchium angustifolium*** BLUE-EYED GRASS | | | |
| flowers | light gray-green | alum | heat |
| | light gray-green | copper | heat |
| | light dull green | iron | decomposition |
| | dark green | tin | heat |
| | light green | none | heat |
| leaves | light bright olive | alum | decomposition |
| | medium khaki-green | copper | decomposition |
| | acid yellow-green | tin | heat |

| DYE SOURCE | DYE RESULT | MORDANT | PROCESSING |
|---|---|---|---|
| ***Solanum dimidiatum*** WESTERN HORSE-NETTLE | | | |
| petals | light olive | iron | heat |
| leaves and stems | light olive | alum | decomposition |
| | light bright olive | copper | heat |
| | light olive | copper | decomposition |
| | light olive | iron | decomposition |
| | acid yellow-green | tin | decomposition |
| | light dull yellow-green | none | heat |
| | light olive | none | decomposition |
| roots | medium olive | iron | heat |
| stamens | light olive | iron | heat |
| ***Solanum elaeagnifolium*** SILVERLEAF NIGHTSHADE | | | |
| leaves | light olive | iron | decomposition |
| ***Solidago missouriensis*** PLAINS GOLDENROD | | | |
| heads | bright medium olive | copper | heat |
| | light olive | iron | decomposition |
| leaves | olive | copper | heat |
| | light olive | iron | decomposition |
| rhizomes and roots | light olive | iron | heat |
| stems | light olive | copper | heat |
| | olive | copper | decomposition |
| | dark olive | iron | heat |
| | olive | iron | decomposition |
| ***Solidago rigida*** STIFF GOLDENROD | | | |
| heads immature | light olive | copper | heat |
| | light olive | iron | decomposition |
| heads mature | light olive | copper | heat |
| | bright olive | iron | decomposition |
| leaves | bright olive | copper | heat |
| | light olive | iron | decomposition |
| roots | light olive | iron | heat |
| stems | light olive | copper | heat |
| | olive | iron | heat |
| ***Sorghastrum nutans*** INDIANGRASS | | | |
| leaves | dark khaki-green | iron | heat |
| | light olive | copper | heat |
| spikelets | dark khaki-green | iron | heat |
| ***Stillingia sylvatica*** QUEEN'S DELIGHT | | | |
| leaves | light olive | copper | heat |
| | medium khaki-green | copper | decomposition |
| pistillate flowers | bright medium olive | copper | heat |
| staminate flowers | olive | copper | heat |
| | light bright olive | copper | decomposition |
| ***Tephrosia virginiana*** GOAT'S RUE | | | |
| leaves | light olive | iron | heat |
| ***Teucrium canadense*** AMERICAN GERMANDER | | | |
| leaves | light bright olive | copper | heat |
| | olive | iron | heat |
| | light dull yellow-green | none | heat |

| DYE SOURCE | DYE RESULT | MORDANT | PROCESSING |
|---|---|---|---|
| ***Ulmus rubra*** SLIPPERY ELM | | | |
| bark | medium khaki-green | iron | heat |
| ***Vernonia baldwinii*** WESTERN IRONWEED | | | |
| flower buds | light olive | copper | heat |
| | dark gray-green | iron | heat |
| | olive | tin | heat |
| leaves | light olive | copper | heat |
| stems | light olive | copper | heat |
| ***Viburnum rufidulum*** RUSTY BLACK HAW | | | |
| leaves | light olive | iron | heat |

# Nature's Most Bountiful Dye Color

Our search for natural dyes resulted in various yellow colors, shown here with Munsell notations. Following these color chips is a list of the plants that were used to obtain these colors, including mordant and processing conditions.

2.5Y 5/6

2.5Y 5/8

2.5Y 6/8

**ANTIQUE GOLD**

5Y 6/8
**ANTIQUE OLIVE·GOLD**

7.5Y 8/4
**BEIGE·YELLOW**

2.5Y 6/10 5Y 7/10 5Y 7/12

**BRIGHT ANTIQUE GOLD**

5Y 6/10

**BRIGHT ANTIQUE OLIVE·GOLD**

2.5Y 7/12 10YR 7/12 10YR 7/14

**BRIGHT GOLD**

1.25Y 8/14 2.5Y 8/12 2.5Y 8/14 2.5Y 8/16

**BRIGHT GOLDEN YELLOW**

10YR 8/12 10YR 8/14

**BRIGHT GOLDEN YELLOW, continued**

10Y 8/10 10Y 8/12

**BRIGHT GREEN·YELLOW**

8.75YR 7/14

**BRIGHT ORANGE·GOLD**

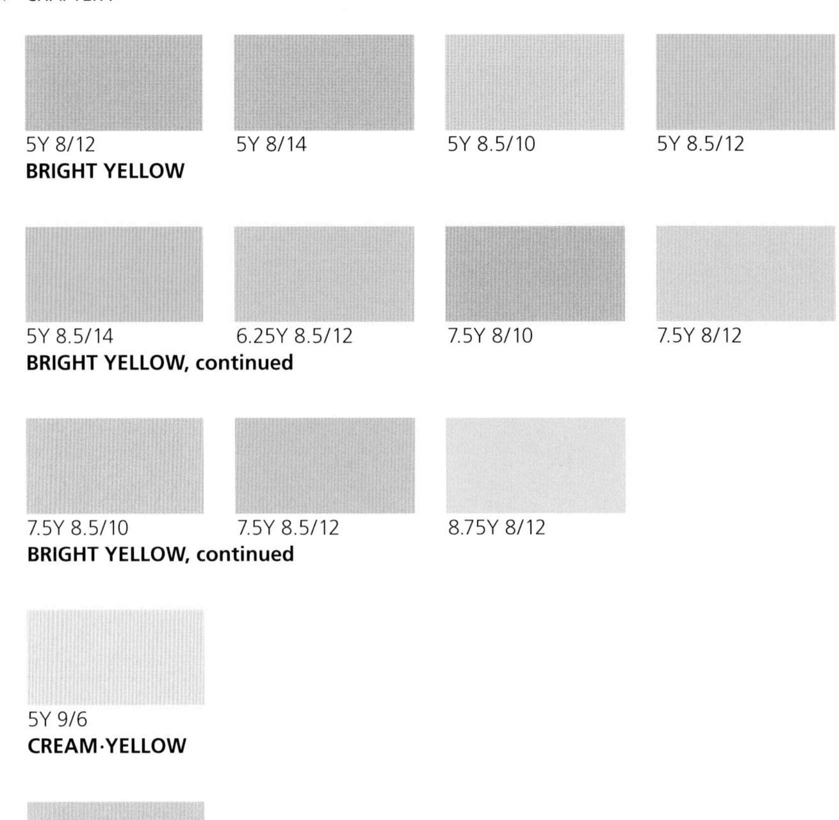

5Y 8/12 5Y 8/14 5Y 8.5/10 5Y 8.5/12

**BRIGHT YELLOW**

5Y 8.5/14 6.25Y 8.5/12 7.5Y 8/10 7.5Y 8/12

**BRIGHT YELLOW, continued**

7.5Y 8.5/10 7.5Y 8.5/12 8.75Y 8/12

**BRIGHT YELLOW, continued**

5Y 9/6

**CREAM·YELLOW**

10Y 8/6

**DULL GREEN·YELLOW**

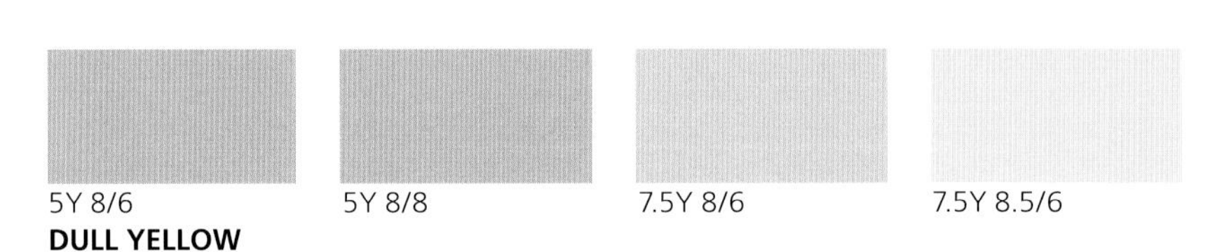

5Y 8/6 5Y 8/8 7.5Y 8/6 7.5Y 8.5/6

**DULL YELLOW**

10Y 8.5/6

**DULL YELLOW, continued**

2.5Y 7/8 2.5Y 7/10 10YR 7/10

**GOLD**

2.5Y 8/8 2.5Y 8/10

**GOLDEN YELLOW**

10Y 8/8

**GREEN·YELLOW**

5Y 7/8

**LIGHT ANTIQUE OLIVE·GOLD**

10Y 8.5/10

**LIGHT BRIGHT GREEN·YELLOW**

10Y 8.5/4

**LIGHT DULL YELLOW**

2.5Y 8.5/10 2.5Y 8.5/12 10YR 8/10

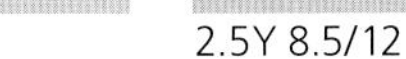

**LIGHT GOLDEN YELLOW**

10Y 8.5/8

**LIGHT GREEN·YELLOW**

5Y 7/6

**LIGHT OLIVE·GOLD**

10YR 8/8

**LIGHT TAN·GOLD**

2.5Y 8.5/8

**LIGHT TAN·YELLOW**

7.5Y 8.5/8 | 7.5Y 9/6 | 10Y 9/6

**LIGHT YELLOW**

7.5Y 9/4 | 10Y 9/4

**PALE YELLOW**

5Y 8/10 | 5Y 8.5/8 | 7.5Y 8/8

**YELLOW**

| DYE SOURCE | DYE RESULT | MORDANT | PROCESSING |
|---|---|---|---|
| ***Acer negundo*** BOXELDER | | | |
| bark | bright golden yellow | tin | heat |
| | cream-yellow | tin | decomposition |
| leaves | bright golden yellow | tin | heat |
| | bright yellow | tin | decomposition |
| ***Acer saccharinum*** SILVER MAPLE | | | |
| leaves | gold | alum | heat |
| | antique gold | copper | heat |
| | bright golden yellow | tin | heat |
| ***Achillea millefolium*** YARROW | | | |
| flowers and phyllaries | light yellow | alum | heat |
| | pale yellow | alum | decomposition |
| | dull yellow | copper | heat |
| | bright yellow | tin | heat |
| | bright yellow | tin | decomposition |
| | pale yellow | none | decomposition |
| leaves | dull yellow | alum | heat |
| | light yellow | alum | decomposition |
| | dull yellow | copper | heat |
| | light dull yellow | copper | decomposition |
| | light golden yellow | tin | heat |
| | bright yellow | tin | decomposition |
| | pale yellow | none | heat |
| | pale yellow | none | decomposition |
| ***Ageratina altissima*** WHITE SNAKEROOT | | | |
| heads | light green-yellow | alum | heat |
| | pale yellow | alum | decomposition |
| | green-yellow | copper | heat |
| | bright golden yellow | tin | heat |
| | bright yellow | tin | decomposition |
| | pale yellow | none | heat |
| | pale yellow | none | decomposition |
| leaves | bright yellow | alum | heat |
| | light olive-gold | copper | heat |
| | bright golden yellow | tin | heat |
| | dull yellow | tin | decomposition |
| | pale yellow | none | heat |
| roots | pale yellow | alum | heat |
| | light dull yellow | copper | heat |
| | pale yellow | tin | heat |
| | pale yellow | tin | decomposition |
| stems | light yellow | alum | heat |
| | pale yellow | alum | decomposition |
| | dull yellow | copper | heat |
| | bright golden yellow | tin | heat |
| | cream-yellow | tin | decomposition |
| ***Alnus maritima*** SEASIDE ALDER | | | |
| leaves | yellow | alum | heat |
| | dull yellow | alum | decomposition |
| | light olive-gold | copper | heat |

| DYE SOURCE | DYE RESULT | MORDANT | PROCESSING |
|---|---|---|---|
| ***Alnus maritima,*** leaves, continued | | | |
| | dull yellow | copper | decomposition |
| | bright golden yellow | tin | heat |
| | light golden yellow | tin | decomposition |
| stems | dull yellow | tin | heat |
| ***Ambrosia psilostachya*** WESTERN RAGWEED | | | |
| leaves and heads | bright yellow | alum | heat |
| | bright yellow | tin | heat |
| | pale yellow | none | heat |
| rhizomes and roots | pale yellow | alum | heat |
| | light dull yellow | copper | heat |
| | beige-yellow | iron | decomposition |
| | pale yellow | tin | heat |
| | light yellow | tin | decomposition |
| | pale yellow | none | heat |
| stems | light yellow | alum | heat |
| | light dull yellow | copper | heat |
| | light yellow | tin | heat |
| | cream-yellow | tin | decomposition |
| | pale yellow | none | heat |
| ***Ambrosia trifida*** GIANT RAGWEED | | | |
| leaves | dull yellow | alum | heat |
| | dull yellow | tin | heat |
| | pale yellow | tin | decomposition |
| roots | pale yellow | tin | heat |
| | pale yellow | tin | decomposition |
| stems | pale yellow | alum | decomposition |
| | beige-yellow | copper | heat |
| | light yellow | copper | decomposition |
| | light yellow | iron | decomposition |
| | dull green-yellow | tin | heat |
| | yellow | tin | decomposition |
| | light yellow | none | decomposition |
| ***Amorpha canescens*** LEADPLANT | | | |
| bark | light golden yellow | tin | heat |
| leaves | yellow | alum | heat |
| | light yellow | alum | decomposition |
| | light olive-gold | copper | heat |
| | pale yellow | copper | decomposition |
| | bright golden yellow | tin | heat |
| | light yellow | tin | decomposition |
| | pale yellow | none | decomposition |
| roots | gold | tin | heat |
| spikes | light yellow | alum | decomposition |
| | bright antique olive-gold | tin | heat |
| ***Apocynum cannabinum*** INDIAN HEMP | | | |
| leaves | bright yellow | alum | heat |
| | light bright green-yellow | alum | decomposition |
| | light olive-gold | copper | heat |
| | dull yellow | copper | decomposition |

| DYE SOURCE | DYE RESULT | MORDANT | PROCESSING |
|---|---|---|---|
| | bright golden yellow | tin | heat |
| | light golden yellow | tin | decomposition |
| | dull yellow | none | heat |
| | pale yellow | none | decomposition |
| roots | pale yellow | alum | decomposition |
| | pale yellow | tin | decomposition |
| ***Argemone polyanthemos*** PRICKLY POPPY | | | |
| leaves | cream-yellow | alum | heat |
| | dull yellow | tin | decomposition |
| pistils | pale yellow | alum | decomposition |
| | light golden yellow | tin | heat |
| | bright yellow | tin | decomposition |
| roots | gold | tin | heat |
| stems | light antique olive-gold | tin | heat |
| | dull yellow | tin | decomposition |
| ***Asclepias arenaria*** SAND MILKWEED | | | |
| flowers | pale yellow | alum | heat |
| | beige-yellow | copper | heat |
| | bright golden yellow | tin | heat |
| fruits | dull yellow | tin | heat |
| | pale yellow | tin | decomposition |
| leaves | light yellow | alum | heat |
| | dull yellow | copper | heat |
| | bright golden yellow | tin | heat |
| | pale yellow | none | heat |
| stems | dull yellow | tin | heat |
| ***Asclepias tuberosa*** BUTTERFLY MILKWEED | | | |
| flowers | light yellow | alum | decomposition |
| | dull yellow | copper | decomposition |
| | yellow | iron | decomposition |
| | bright antique gold | tin | heat |
| | bright golden yellow | tin | decomposition |
| | pale yellow | none | decomposition |
| leaves | light yellow | alum | heat |
| | dull yellow | copper | heat |
| | pale yellow | iron | decomposition |
| | bright golden yellow | tin | heat |
| | cream-yellow | tin | decomposition |
| | pale yellow | none | heat |
| ***Asclepias viridis*** ANTELOPE-HORN MILKWEED | | | |
| flowers | pale yellow | alum | decomposition |
| | pale yellow | copper | decomposition |
| | yellow | tin | heat |
| | light yellow | tin | decomposition |
| leaves | dull yellow | alum | heat |
| | light bright green-yellow | alum | decomposition |
| | light olive-gold | copper | decomposition |
| | antique gold | iron | heat |
| | antique gold | iron | decomposition |
| | bright golden yellow | tin | heat |

| DYE SOURCE | DYE RESULT | MORDANT | PROCESSING |
|---|---|---|---|
| ***Asclepias viridis,*** leaves, continued | | | |
| | bright golden yellow | tin | decomposition |
| | pale yellow | none | heat |
| | pale yellow | none | decomposition |
| ***Asimina triloba*** PAWPAW | | | |
| branches young | cream-yellow | tin | heat |
| leaves | dull yellow | alum | heat |
| | bright golden yellow | tin | heat |
| ***Baccharis salicina*** WILLOW BACCHARIS | | | |
| leaves | bright yellow | alum | heat |
| | pale yellow | alum | decomposition |
| | dull yellow | copper | heat |
| | bright golden yellow | tin | heat |
| | light yellow | tin | decomposition |
| | pale yellow | none | heat |
| stems | bright golden yellow | tin | heat |
| | cream yellow | tin | decomposition |
| ***Baptisia australis*** BLUE WILD-INDIGO | | | |
| petals | yellow | tin | heat |
| | dull yellow | tin | decomposition |
| roots | gold | tin | heat |
| stems (basal) | light antique olive-gold | tin | heat |
| ***Baptisia bracteata*** PLAINS WILD-INDIGO | | | |
| leaves | bright yellow | alum | heat |
| | bright yellow | tin | heat |
| | pale yellow | none | heat |
| roots | golden yellow | tin | heat |
| | pale yellow | tin | decomposition |
| ***Baptisia sphaerocarpa*** GOLDEN WILD-INDIGO | | | |
| leaves | light yellow | tin | heat |
| petals | pale yellow | alum | heat |
| | dull yellow | iron | heat |
| | light yellow | tin | heat |
| roots | golden yellow | tin | heat |
| stems | dull yellow | tin | heat |
| ***Bidens aristosa*** BEARDED BEGGAR-TICKS | | | |
| flowers | antique gold | alum | heat |
| | antique gold | copper | heat |
| | antique gold | none | heat |
| leaves | gold | alum | heat |
| | antique gold | copper | heat |
| | light olive-gold | none | heat |
| stems | dull yellow | copper | heat |
| | bright gold | tin | heat |
| ***Bidens bipinnata*** SPANISH NEEDLES | | | |
| heads | bright golden yellow | tin | heat |
| leaves | antique gold | copper | heat |
| roots | light yellow | alum | heat |
| | cream-yellow | tin | heat |

| DYE SOURCE | DYE RESULT | MORDANT | PROCESSING |
|---|---|---|---|
| stems | light yellow | alum | heat |
| | bright golden yellow | tin | heat |
| | light yellow | none | heat |
| ***Brickellia eupatorioides*** FALSE BONESET | | | |
| heads | bright yellow | alum | heat |
| | light yellow | alum | decomposition |
| | dull yellow | copper | decomposition |
| | dull yellow | iron | decomposition |
| | bright golden yellow | tin | heat |
| | bright yellow | tin | decomposition |
| | dull yellow | none | heat |
| | pale yellow | none | decomposition |
| leaves | yellow | alum | heat |
| | light yellow | alum | decomposition |
| | dull yellow | iron | decomposition |
| | bright golden yellow | tin | heat |
| | light yellow | tin | decomposition |
| | light yellow | none | heat |
| | pale yellow | none | decomposition |
| roots | dull yellow | alum | heat |
| | bright yellow | tin | heat |
| | pale yellow | none | heat |
| stems | pale yellow | alum | heat |
| | beige-yellow | copper | heat |
| | dull yellow | iron | decomposition |
| | light yellow | tin | heat |
| | dull yellow | tin | decomposition |
| ***Bumelia lanuginosa*** CHITTAMWOOD | | | |
| branchlets | cream-yellow | tin | heat |
| fruits | pale yellow | alum | decomposition |
| | beige-yellow | iron | decomposition |
| | light golden yellow | tin | decomposition |
| leaves | light yellow | alum | heat |
| | light yellow | alum | decomposition |
| | dull green-yellow | copper | heat |
| | light golden yellow | tin | heat |
| | cream-yellow | tin | decomposition |
| | pale yellow | none | heat |
| ***Callicarpa americana*** AMERICAN BEAUTYBERRY | | | |
| branchlets | light yellow | alum | heat |
| | dull yellow | copper | heat |
| | light bright green-yellow | tin | heat |
| | dull yellow | tin | decomposition |
| | pale yellow | none | heat |
| fruits | light yellow | alum | heat |
| | pale yellow | copper | heat |
| | light yellow | tin | heat |
| | dull yellow | tin | decomposition |
| leaves | yellow | alum | heat |
| | dull yellow | copper | heat |

| DYE SOURCE | DYE RESULT | MORDANT | PROCESSING |
|---|---|---|---|
| ***Calliicarpa americana,*** leaves, continued | | | |
| | yellow | tin | heat |
| | dull yellow | tin | decomposition |
| | pale yellow | none | heat |
| ***Calylophus hartwegii*** HARTWEG'S SUNDROP | | | |
| leaves | dull yellow | alum | heat |
| | bright golden yellow | alum | decomposition |
| | dull yellow | copper | heat |
| | bright golden yellow | tin | decomposition |
| petals | light golden yellow | tin | heat |
| | golden yellow | tin | decomposition |
| ***Calylophus serrulatus*** HALFSHRUB SUNDROP | | | |
| leaves | dull yellow | alum | heat |
| | dull yellow | alum | decomposition |
| | bright golden yellow | tin | heat |
| | bright golden yellow | tin | decomposition |
| petals | light golden yellow | tin | heat |
| stems | bright golden yellow | tin | heat |
| | light yellow | tin | decomposition |
| ***Campsis radicans*** TRUMPET CREEPER | | | |
| calyx | pale yellow | alum | heat |
| | dull yellow | tin | heat |
| flowers | pale yellow | alum | heat |
| | pale yellow | alum | decomposition |
| | pale yellow | copper | decomposition |
| | light olive-gold | iron | heat |
| | pale yellow | iron | decomposition |
| | pale yellow | tin | decomposition |
| | pale yellow | none | decomposition |
| fruits | light yellow | tin | heat |
| leaves | bright yellow | alum | heat |
| | light olive-gold | copper | heat |
| | bright yellow | tin | heat |
| | pale yellow | none | heat |
| ***Carya cordiformis*** BITTERNUT HICKORY | | | |
| branches | light tan-gold | tin | heat |
| fruits | light yellow | alum | heat |
| | beige-yellow | copper | heat |
| | light yellow | tin | heat |
| leaves | light yellow | alum | heat |
| | beige-yellow | copper | heat |
| | light golden yellow | tin | heat |
| ***Carya illinoinensis*** PECAN | | | |
| branchlets | light olive-gold | iron | decomposition |
| | dull yellow | tin | decomposition |
| ***Castanea ozarkensis*** OZARK CHINKAPIN | | | |
| leaves | pale yellow | alum | heat |
| | pale yellow | alum | decomposition |
| | beige-yellow | copper | heat |
| | dull yellow | tin | heat |
| | bright golden yellow | tin | decomposition |

| DYE SOURCE | DYE RESULT | MORDANT | PROCESSING |
|---|---|---|---|
| ***Castilleja indivisa*** INDIAN PAINTBRUSH | | | |
| bract tips | dull yellow | tin | decomposition |
| leaves and stems | gold | copper | decomposition |
| | bright antique gold | tin | decomposition |
| | light dull yellow | none | heat |
| ***Castilleja purpurea* var. *citrina*** YELLOW PAINTBRUSH | | | |
| leaves | gold | iron | heat |
| | bright yellow | tin | heat |
| | cream-yellow | tin | decomposition |
| petals and bracts | dull yellow | alum | heat |
| | dull yellow | copper | heat |
| | yellow | tin | heat |
| | cream-yellow | tin | decomposition |
| ***Catalpa speciosa*** NORTHERN CATALPA | | | |
| branches | golden yellow | alum | decomposition |
| | gold | copper | decomposition |
| | cream-yellow | tin | decomposition |
| | gold | none | decomposition |
| branchlets | antique gold | tin | heat |
| leaves | light olive-gold | none | decomposition |
| | antique gold | tin | heat |
| | bright yellow | tin | decomposition |
| ***Ceanothus americanus*** NEW JERSEY TEA | | | |
| flowers | light yellow | alum | heat |
| | dull yellow | copper | heat |
| | bright golden yellow | tin | heat |
| | pale yellow | tin | decomposition |
| leaves | light yellow | alum | heat |
| | pale yellow | alum | decomposition |
| | bright golden yellow | tin | heat |
| | light golden yellow | tin | decomposition |
| | pale yellow | none | decomposition |
| stems | light golden yellow | tin | heat |
| ***Centaurea americana*** BASKET FLOWER | | | |
| disk florets | light dull yellow | copper | heat |
| | pale yellow | tin | heat |
| leaves and stems | gold | tin | heat |
| | dull yellow | none | heat |
| ***Cephalanthus occidentalis*** BUTTONBUSH | | | |
| flowers | bright golden yellow | tin | heat |
| | yellow | tin | decomposition |
| leaves | bright golden yellow | tin | heat |
| | yellow | tin | decomposition |
| ***Cercis canadensis*** REDBUD | | | |
| bark of branches | bright golden yellow | tin | heat |
| flowers | dull yellow | alum | decomposition |
| | yellow | copper | decomposition |
| | golden yellow | iron | decomposition |
| | antique gold | tin | heat |
| | yellow | none | decomposition |

| DYE SOURCE | DYE RESULT | MORDANT | PROCESSING |
|---|---|---|---|
| ***Chrysopsis pilosa*** SOFT GOLDEN-ASTER | | | |
| leaves | bright yellow | alum | heat |
| | pale yellow | alum | decomposition |
| | bright golden yellow | tin | heat |
| | light yellow | tin | decomposition |
| | light dull yellow | none | heat |
| petals | pale yellow | alum | decomposition |
| | light dull yellow | copper | decomposition |
| | cream-yellow | tin | decomposition |
| roots | dull yellow | alum | heat |
| | light yellow | tin | heat |
| ***Cicuta maculata*** WATER-HEMLOCK | | | |
| flowers | bright golden yellow | alum | heat |
| | pale yellow | alum | decomposition |
| | gold | copper | heat |
| | bright yellow | tin | decomposition |
| leaves | light yellow | alum | decomposition |
| | light antique olive-gold | copper | heat |
| | beige-yellow | iron | decomposition |
| | light golden yellow | tin | decomposition |
| | dull yellow | none | heat |
| roots | beige-yellow | iron | heat |
| stems | pale yellow | copper | heat |
| | yellow | tin | heat |
| ***Cirsium undulatum*** WAVY-LEAF THISTLE | | | |
| pappus | light olive-gold | iron | heat |
| | pale yellow | iron | decomposition |
| | dull yellow | tin | heat |
| | light yellow | tin | decomposition |
| leaves | bright green-yellow | alum | heat |
| | antique olive-gold | tin | heat |
| ***Comandra umbellata*** BASTARD TOAD-FLAX | | | |
| leaves | yellow | alum | heat |
| | gold | copper | decomposition |
| | light tan-gold | tin | decomposition |
| stems | yellow | alum | heat |
| ***Coreopsis tinctoria*** PLAINS TICKSEED | | | |
| disk florets and phyllaries | bright antique gold | alum | heat |
| | bright gold | tin | decomposition |
| | antique gold | none | heat |
| leaves | bright gold | alum | heat |
| | yellow | alum | decomposition |
| | antique gold | copper | heat |
| | light olive-gold | copper | decomposition |
| | bright gold | tin | decomposition |
| | light olive-gold | none | heat |
| | dull yellow | none | decomposition |
| petals | gold | copper | decomposition |
| | cream-yellow | none | decomposition |
| ray florets | antique olive-gold | alum | heat |
| | antique olive-gold | none | heat |

| DYE SOURCE | DYE RESULT | MORDANT | PROCESSING |
|---|---|---|---|
| ***Cornus drummondii*** ROUGH-LEAF DOGWOOD | | | |
| fruits | pale yellow | alum | heat |
| | dull yellow | alum | decomposition |
| | light olive-gold | iron | decomposition |
| | light golden yellow | tin | heat |
| | bright golden yellow | tin | decomposition |
| leaves | dull yellow | alum | heat |
| | light tan-gold | alum | decomposition |
| | light golden yellow | tin | heat |
| pedicels | light dull yellow | alum | heat |
| | beige-yellow | iron | decomposition |
| | yellow | tin | heat |
| | light tan-yellow | tin | decomposition |
| roots | dull yellow | alum | heat |
| | pale yellow | alum | decomposition |
| | beige-yellow | copper | decomposition |
| | pale yellow | tin | decomposition |
| ***Cornus florida*** FLOWERING DOGWOOD | | | |
| bark | golden yellow | tin | heat |
| fruits | pale yellow | alum | heat |
| | yellow | tin | heat |
| | golden yellow | tin | decomposition |
| leaves | yellow | tin | heat |
| ***Cotinus obovatus*** SMOKE-TREE | | | |
| bark | gold | alum | heat |
| | bright antique gold | copper | heat |
| | dull yellow | none | heat |
| leaves | dull yellow | alum | heat |
| | bright golden yellow | tin | heat |
| | light tan-gold | tin | decomposition |
| wood | yellow | alum | decomposition |
| | dull yellow | copper | decomposition |
| | light olive-gold | iron | decomposition |
| | bright antique gold | tin | heat |
| | bright golden yellow | tin | decomposition |
| | yellow | none | decomposition |
| ***Cucurbita foetidissima*** BUFFALO GOURD | | | |
| fruits | dull yellow | tin | heat |
| leaves | cream-yellow | alum | decomposition |
| stems | yellow | tin | heat |
| ***Cycloloma atriplicifolium*** TUMBLE RINGWING | | | |
| leaves | light yellow | alum | heat |
| | dull yellow | copper | heat |
| | bright golden yellow | tin | heat |
| | light yellow | tin | decomposition |
| stems | light yellow | alum | heat |
| | dull yellow | copper | heat |
| | bright golden yellow | tin | heat |
| | light yellow | tin | decomposition |
| | pale yellow | none | heat |

| DYE SOURCE | DYE RESULT | MORDANT | PROCESSING |
|---|---|---|---|
| ***Cyperus squarrosus*** BEARDED FLATSEDGE | | | |
| leaves | pale yellow | alum | decomposition |
| | cream-yellow | tin | decomposition |
| spikes | bright golden yellow | tin | heat |
| | cream-yellow | tin | decomposition |
| ***Dalea candida*** WHITE PRAIRIE-CLOVER | | | |
| flowers | yellow | alum | decomposition |
| | bright gold | tin | heat |
| | bright golden yellow | tin | decomposition |
| leaves | light yellow | alum | heat |
| | light green-yellow | alum | decomposition |
| | dull yellow | copper | heat |
| | light tan-yellow | iron | decomposition |
| | bright gold | tin | heat |
| | pale yellow | none | heat |
| | pale yellow | none | decomposition |
| roots | beige-yellow | iron | decomposition |
| | pale yellow | tin | decomposition |
| ***Dalea lanata*** WOOLLY DALEA | | | |
| flowers | bright yellow | alum | heat |
| | dull yellow | alum | decomposition |
| | bright antique gold | copper | heat |
| | light olive-gold | copper | decomposition |
| | bright golden yellow | tin | heat |
| | bright yellow | none | heat |
| | dull yellow | none | decomposition |
| leaves | bright yellow | alum | heat |
| | bright yellow | alum | decomposition |
| | gold | copper | heat |
| | bright antique olive-gold | iron | heat |
| | light yellow | tin | decomposition |
| | bright yellow | none | heat |
| | light yellow | none | decomposition |
| roots | cream-yellow | alum | heat |
| | yellow | tin | heat |
| stems | light yellow | alum | heat |
| | dull yellow | copper | heat |
| | dull yellow | iron | heat |
| | light yellow | tin | heat |
| | pale yellow | tin | decomposition |
| | pale yellow | none | heat |
| ***Dalea purpurea*** PURPLE PRAIRIE-CLOVER | | | |
| flowers | dull yellow | iron | heat |
| | dull yellow | tin | heat |
| leaves and stems | light yellow | alum | heat |
| | bright yellow | alum | decomposition |
| | dull yellow | copper | heat |
| | bright golden yellow | tin | heat |
| | bright golden yellow | tin | decomposition |

| DYE SOURCE | DYE RESULT | MORDANT | PROCESSING |
|---|---|---|---|
| ***Dalea villosa*** SILKY PRAIRIE-CLOVER | | | |
| flowers | light yellow | alum | heat |
| | light yellow | alum | decomposition |
| | dull yellow | copper | heat |
| | bright golden yellow | tin | heat |
| | light golden yellow | tin | decomposition |
| leaves | bright yellow | alum | heat |
| | bright yellow | alum | decomposition |
| | light antique olive-gold | copper | heat |
| | beige-yellow | copper | decomposition |
| | bright golden yellow | tin | heat |
| | light golden yellow | tin | decomposition |
| | pale yellow | none | heat |
| roots | dull yellow | alum | heat |
| | pale yellow | alum | decomposition |
| | dull yellow | copper | heat |
| | dull yellow | iron | heat |
| | yellow | tin | heat |
| | dull yellow | none | heat |
| stems | light yellow | alum | heat |
| | pale yellow | alum | decomposition |
| | bright golden yellow | tin | heat |
| | cream-yellow | tin | decomposition |
| ***Desmanthus illinoensis*** ILLINOIS BUNDLEFLOWER | | | |
| fruits | light olive-gold | alum | heat |
| | bright golden yellow | tin | heat |
| | cream-yellow | tin | decomposition |
| leaves | bright yellow | alum | heat |
| | pale yellow | alum | decomposition |
| | bright golden yellow | tin | heat |
| | pale yellow | tin | decomposition |
| stems | light yellow | alum | heat |
| | pale yellow | alum | decomposition |
| | dull yellow | iron | decomposition |
| | bright yellow | tin | heat |
| ***Desmodium glutinosum*** STICKY TICKCLOVER | | | |
| leaves | cream-yellow | alum | heat |
| | yellow | tin | heat |
| ***Diospyros virginiana*** PERSIMMON | | | |
| branchlets | golden yellow | alum | heat |
| | light olive-gold | copper | heat |
| fruits | pale yellow | alum | heat |
| | pale yellow | alum | decomposition |
| | light dull yellow | copper | heat |
| | beige yellow | iron | decomposition |
| | cream-yellow | tin | heat |
| | cream-yellow | tin | decomposition |
| leaves | gold | alum | heat |
| | antique gold | iron | heat |
| | bright gold | tin | heat |

| DYE SOURCE | DYE RESULT | MORDANT | PROCESSING |
|---|---|---|---|
| ***Diospyros virginiana,*** leaves, continued | | | |
| | light tan-yellow | tin | decomposition |
| | gold | none | heat |
| ***Dracopis amplexicaulis*** CLASPING-LEAVED CONEFLOWER | | | |
| disk florets | light antique olive-gold | tin | decomposition |
| leaves | light yellow | alum | heat |
| | light olive-gold | copper | heat |
| | bright golden yellow | tin | heat |
| | pale yellow | none | heat |
| ray florets | antique olive-gold | tin | heat |
| | yellow | tin | decomposition |
| roots | beige-yellow | copper | heat |
| | cream-yellow | tin | heat |
| stems | yellow | tin | heat |
| ***Echinacea angustifolia*** PURPLE PRAIRIE CONEFLOWER | | | |
| disk florets | light antique olive-gold | tin | heat |
| leaves | dull yellow | alum | heat |
| | light dull yellow | alum | decomposition |
| | gold | tin | heat |
| | dull yellow | tin | decomposition |
| ray florets | cream-yellow | alum | heat |
| | bright golden yellow | tin | heat |
| roots | dull yellow | iron | heat |
| | cream-yellow | tin | heat |
| ***Echinocereus reichenbachii*** LACE HEDGEHOG-CACTUS | | | |
| flowers | light yellow | alum | heat |
| | beige-yellow | iron | decomposition |
| | bright orange-gold | tin | heat |
| | bright yellow | tin | decomposition |
| ***Elephantopus carolinianus*** LEAFY ELEPHANT'S FOOT | | | |
| leaves | dull yellow | alum | heat |
| | beige-yellow | iron | heat |
| | dull yellow | tin | heat |
| | light yellow | tin | decomposition |
| roots | light yellow | alum | heat |
| | light yellow | tin | heat |
| | light yellow | tin | decomposition |
| stems | light yellow | alum | heat |
| | light dull yellow | copper | heat |
| | cream-yellow | tin | heat |
| ***Erigeron strigosus*** DAISY FLEABANE | | | |
| disk florets | dull yellow | alum | heat |
| | light olive-gold | copper | heat |
| | bright golden yellow | tin | heat |
| | dull yellow | none | heat |
| ***Euphorbia maculata*** SPOTTED SPURGE | | | |
| leaves and stems | dull yellow | alum | heat |
| | light yellow | tin | heat |
| | pale yellow | tin | decomposition |

| DYE SOURCE | DYE RESULT | MORDANT | PROCESSING |
|---|---|---|---|
| ***Euphorbia marginata*** SNOW-ON-THE-MOUNTAIN | | | |
| bracts | pale yellow | alum | heat |
| | pale yellow | iron | heat |
| | pale yellow | tin | decomposition |
| | pale yellow | none | heat |
| | pale yellow | none | decomposition |
| cyathia | light yellow | alum | heat |
| | light yellow | alum | decomposition |
| | dull yellow | copper | heat |
| | dull yellow | iron | heat |
| | bright yellow | tin | heat |
| | light yellow | tin | decomposition |
| | pale yellow | none | heat |
| | pale yellow | none | decomposition |
| leaves | light yellow | alum | heat |
| | green-yellow | copper | heat |
| | bright golden yellow | tin | heat |
| | cream-yellow | tin | decomposition |
| | pale yellow | none | heat |
| roots | dull yellow | tin | heat |
| | pale yellow | tin | decomposition |
| stems | pale yellow | alum | heat |
| ***Eustoma exaltatum*** PRAIRIE GENTIAN | | | |
| leaves | golden yellow | tin | heat |
| petals | pale yellow | alum | decomposition |
| | dull yellow | tin | heat |
| | dull yellow | tin | heat |
| | bright yellow | tin | decomposition |
| roots | light tan-yellow | tin | heat |
| stems | light tan-yellow | tin | heat |
| ***Fraxinus americana*** WHITE ASH | | | |
| fruits | pale yellow | alum | heat |
| | light dull yellow | copper | heat |
| | light dull yellow | iron | heat |
| | light yellow | tin | heat |
| | dull yellow | tin | decomposition |
| ***Fraxinus pennsylvanica*** GREEN ASH | | | |
| bark | beige-yellow | iron | decomposition |
| | bright yellow | tin | heat |
| | light yellow | tin | decomposition |
| leaves | bright yellow | alum | heat |
| | pale yellow | copper | decomposition |
| | beige-yellow | iron | decomposition |
| | bright antique gold | tin | heat |
| | light yellow | tin | decomposition |
| ***Fraxinus quadrangulata*** BLUE ASH | | | |
| bark | dull yellow | alum | heat |
| | pale yellow | alum | decomposition |
| | gold | tin | heat |
| | pale yellow | tin | decomposition |

| DYE SOURCE | DYE RESULT | MORDANT | PROCESSING |
|---|---|---|---|
| ***Fraxinus quadrangulata,*** continued | | | |
| leaves | light yellow | alum | heat |
| | pale yellow | alum | decomposition |
| | light olive-gold | copper | heat |
| | gold | iron | heat |
| | bright golden yellow | tin | heat |
| | light bright green-yellow | tin | decomposition |
| ***Froelichia floridana*** FIELD SNAKE-COTTON | | | |
| flowers | light yellow | alum | heat |
| | bright golden yellow | tin | heat |
| leaves | light yellow | alum | heat |
| | bright golden yellow | tin | heat |
| stems | light yellow | alum | heat |
| | bright golden yellow | tin | heat |
| | pale yellow | tin | decomposition |
| ***Gaillardia aestivalis*** PRAIRIE GAILLARDIA | | | |
| disk florets | dull yellow | alum | decomposition |
| | dull yellow | copper | decomposition |
| | yellow | tin | decomposition |
| | pale yellow | none | decomposition |
| leaves | dull yellow | alum | decomposition |
| | bright antique gold | tin | heat |
| | light dull yellow | tin | decomposition |
| | dull yellow | none | heat |
| ray florets | light dull yellow | alum | heat |
| roots | dull yellow | tin | heat |
| ***Gaillardia pulchella*** INDIAN BLANKET | | | |
| disk florets | light olive-gold | iron | heat |
| | bright gold | tin | heat |
| | yellow | tin | decomposition |
| roots | cream-yellow | alum | decomposition |
| ***Glandularia canadensis*** ROSE VERVAIN | | | |
| flowers | pale yellow | alum | heat |
| | dull yellow | copper | heat |
| | light antique olive-gold | tin | decomposition |
| leaves and stems | pale yellow | alum | heat |
| | light dull yellow | copper | heat |
| ***Gleditsia triacanthos*** HONEY LOCUST | | | |
| fruits | pale yellow | copper | heat |
| | yellow | tin | heat |
| | cream-yellow | tin | decomposition |
| ***Gymnocladus dioicus*** KENTUCKY COFFEE-TREE | | | |
| bark | cream-yellow | alum | heat |
| | pale yellow | alum | decomposition |
| | dull yellow | copper | heat |
| | beige-yellow | copper | decomposition |
| | bright golden yellow | tin | heat |
| | pale yellow | tin | decomposition |
| fruits | pale yellow | alum | decomposition |
| | pale yellow | copper | heat |
| | yellow | tin | heat |

| DYE SOURCE | DYE RESULT | MORDANT | PROCESSING |
|---|---|---|---|
| | cream-yellow | tin | decomposition |
| leaflets | light yellow | alum | heat |
| | pale yellow | alum | decomposition |
| | dull yellow | copper | heat |
| | light golden yellow | tin | heat |
| | yellow | tin | decomposition |
| | pale yellow | none | heat |
| | pale yellow | none | decomposition |
| petioles | pale yellow | alum | heat |
| | light dull yellow | copper | heat |
| | light dull yellow | iron | heat |
| | bright yellow | tin | heat |
| | pale yellow | tin | decomposition |
| | pale yellow | none | heat |
| seeds | pale yellow | alum | decomposition |
| | light dull yellow | iron | heat |
| | light yellow | tin | heat |
| | pale yellow | tin | decomposition |
| | pale yellow | none | decomposition |
| stems | cream-yellow | tin | heat |
| ***Hedyotis nigricans*** PRAIRIE BLUETS | | | |
| leaves and stems | yellow | tin | decomposition |
| ***Helenium amarum*** BITTER SNEEZEWEED | | | |
| heads | light yellow | alum | heat |
| | dull green-yellow | copper | heat |
| | light olive-gold | iron | decomposition |
| | bright yellow | tin | heat |
| | dull yellow | tin | decomposition |
| | pale yellow | none | heat |
| roots | pale yellow | alum | heat |
| | pale yellow | copper | heat |
| | light dull yellow | iron | heat |
| | pale yellow | tin | heat |
| | pale yellow | tin | decomposition |
| | pale yellow | none | heat |
| stems | pale yellow | alum | heat |
| | dull yellow | iron | decomposition |
| | light yellow | tin | heat |
| ***Helianthus annuus*** ANNUAL SUNFLOWER | | | |
| disk florets | beige-yellow | alum | heat |
| | bright antique gold | tin | heat |
| | dull yellow | tin | decomposition |
| leaves | yellow | tin | heat |
| | cream-yellow | tin | decomposition |
| ray florets | pale yellow | alum | heat |
| | pale yellow | alum | decomposition |
| | pale yellow | copper | decomposition |
| | pale yellow | iron | decomposition |
| | bright golden yellow | tin | heat |
| | light yellow | tin | decomposition |
| | pale yellow | none | decomposition |

| DYE SOURCE | DYE RESULT | MORDANT | PROCESSING |
|---|---|---|---|
| ***Helianthus maximiliani*** MAXIMILIAN'S SUNFLOWER | | | |
| disk florets and phyllaries | bright yellow | alum | heat |
| | cream-yellow | alum | decomposition |
| | gold | copper | heat |
| | light golden yellow | tin | heat |
| | yellow | tin | decomposition |
| | light yellow | none | heat |
| leaves | bright yellow | alum | heat |
| | dull yellow | alum | decomposition |
| | bright antique gold | copper | heat |
| | bright yellow | tin | heat |
| | light yellow | tin | decomposition |
| | light yellow | none | heat |
| ray florets | yellow | alum | heat |
| | pale yellow | alum | decomposition |
| | dull yellow | copper | heat |
| | dull yellow | copper | decomposition |
| | dull yellow | iron | decomposition |
| | bright golden yellow | tin | heat |
| | bright yellow | tin | decomposition |
| | dull yellow | none | heat |
| rhizomes | light yellow | alum | heat |
| | light dull yellow | copper | heat |
| | light yellow | tin | heat |
| | dull yellow | tin | decomposition |
| | pale yellow | none | heat |
| stems | light yellow | alum | heat |
| | yellow | copper | heat |
| | light olive-gold | iron | decomposition |
| | bright yellow | tin | heat |
| | light yellow | tin | decomposition |
| | cream-yellow | none | heat |
| ***Helianthus mollis*** ASHY SUNFLOWER | | | |
| disk florets | yellow | alum | heat |
| | dull yellow | tin | heat |
| | pale yellow | none | heat |
| leaves | light antique olive-gold | alum | heat |
| | pale yellow | alum | decomposition |
| | pale yellow | copper | decomposition |
| | pale yellow | iron | decomposition |
| | yellow | tin | heat |
| | yellow | tin | decomposition |
| | light olive-gold | none | heat |
| | pale yellow | none | decomposition |
| ray florets | beige-yellow | alum | heat |
| | yellow | tin | heat |
| ***Helianthus tuberosus*** JERUSALEM ARTICHOKE | | | |
| disk and ray florets | bright yellow | alum | heat |
| | pale yellow | alum | decomposition |
| | bright yellow | tin | decomposition |
| | light yellow | none | heat |

| DYE SOURCE | DYE RESULT | MORDANT | PROCESSING |
|---|---|---|---|
| | pale yellow | none | decomposition |
| leaves | bright yellow | alum | heat |
| | bright yellow | tin | heat |
| | pale yellow | none | heat |
| roots | pale yellow | alum | heat |
| | pale yellow | tin | heat |
| stems | light yellow | alum | heat |
| | bright yellow | tin | heat |
| | light yellow | tin | decomposition |
| ***Hibiscus laevis*** SCARLET ROSE MALLOW | | | |
| leaves | light yellow | alum | heat |
| | bright golden yellow | tin | heat |
| | light tan-yellow | tin | decomposition |
| petals | bright yellow | alum | heat |
| | pale yellow | alum | decomposition |
| | antique gold | copper | heat |
| | bright golden yellow | tin | heat |
| | light golden yellow | tin | decomposition |
| sepals | cream-yellow | tin | heat |
| stems | pale yellow | alum | heat |
| | beige-yellow | copper | heat |
| | light golden yellow | tin | heat |
| ***Hydrangea arborescens*** WILD HYDRANGEA | | | |
| flowers | light yellow | alum | heat |
| | pale yellow | alum | decomposition |
| | dull yellow | copper | heat |
| | bright golden yellow | tin | heat |
| | yellow | tin | decomposition |
| leaves | bright golden yellow | tin | heat |
| | cream-yellow | tin | decomposition |
| ***Ipomoea leptophylla*** BUSH MORNING-GLORY | | | |
| leaves | yellow | alum | heat |
| | dull yellow | copper | heat |
| | bright golden yellow | tin | heat |
| rootstocks | light tan-yellow | tin | heat |
| stems | yellow | tin | heat |
| | yellow | tin | decomposition |
| ***Ipomopsis rubra*** STANDING CYPRESS | | | |
| leaves | light yellow | alum | heat |
| | light yellow | alum | decomposition |
| | dull yellow | copper | heat |
| | beige-yellow | copper | decomposition |
| | bright golden yellow | tin | decomposition |
| | pale yellow | none | heat |
| | light yellow | none | decomposition |
| petals | dull yellow | copper | decomposition |
| | light olive-gold | iron | decomposition |
| | bright yellow | tin | heat |
| | bright yellow | tin | decomposition |
| roots | cream-yellow | tin | heat |
| | light yellow | tin | decomposition |

| DYE SOURCE | DYE RESULT | MORDANT | PROCESSING |
|---|---|---|---|
| ***Ipomopsis rubra,*** continued | | | |
| stems | pale yellow | alum | heat |
| | bright golden yellow | tin | heat |
| | cream-yellow | tin | decomposition |
| ***Juniperus virginiana*** EASTERN RED CEDAR | | | |
| leaves and stems | light yellow | alum | heat |
| | pale yellow | alum | decomposition |
| | dull yellow | copper | heat |
| | dull yellow | copper | decomposition |
| | dull yellow | iron | decomposition |
| | bright yellow | tin | heat |
| | light tan-yellow | tin | decomposition |
| | pale yellow | none | heat |
| ***Lesquerella gracilis*** SPREADING BLADDERPOD | | | |
| leaves and stems | dull yellow | iron | heat |
| | light golden yellow | tin | heat |
| ***Liatris punctata*** DOTTED GAYFEATHER | | | |
| disk florets | pale yellow | alum | decomposition |
| | pale yellow | copper | heat |
| | light dull yellow | tin | heat |
| | light yellow | tin | decomposition |
| | pale yellow | none | decomposition |
| leaves and stems | pale yellow | alum | heat |
| | light golden yellow | tin | heat |
| | bright yellow | tin | decomposition |
| phyllaries | light golden yellow | tin | heat |
| | dull yellow | tin | decomposition |
| roots | pale yellow | copper | decomposition |
| | dull yellow | iron | heat |
| | pale yellow | tin | decomposition |
| ***Liatris squarrosa*** SCALY GAYFEATHER | | | |
| flowers | beige-yellow | copper | heat |
| leaves | light yellow | alum | heat |
| | light yellow | alum | decomposition |
| | bright gold | tin | heat |
| | cream-yellow | tin | decomposition |
| | pale yellow | none | heat |
| roots | cream-yellow | tin | heat |
| ***Linum sulcatum*** GROOVED FLAX | | | |
| leaves, stems, and petals | yellow | alum | heat |
| | light yellow | alum | decomposition |
| | light antique olive-gold | copper | heat |
| | dull yellow | copper | decomposition |
| | light yellow | tin | decomposition |
| | pale yellow | none | decomposition |
| roots | dull yellow | iron | heat |
| ***Liquidambar styraciflua*** SWEET GUM | | | |
| leaves | golden yellow | alum | decomposition |
| roots | pale yellow | alum | heat |
| | pale yellow | tin | heat |

| DYE SOURCE | DYE RESULT | MORDANT | PROCESSING |
|---|---|---|---|
| ***Maclura pomifera*** OSAGE ORANGE | | | |
| bark | golden yellow | alum | heat |
| | pale yellow | alum | decomposition |
| | light golden yellow | tin | heat |
| | pale yellow | tin | decomposition |
| fruits | golden yellow | alum | heat |
| | light olive-gold | copper | heat |
| | bright golden yellow | tin | heat |
| | dull yellow | none | heat |
| leaves | bright yellow | alum | heat |
| | light olive-gold | copper | heat |
| | bright gold | tin | heat |
| | cream-yellow | tin | decomposition |
| roots | pale yellow | alum | decomposition |
| | bright antique gold | copper | heat |
| | light dull yellow | copper | decomposition |
| | bright gold | tin | heat |
| | light yellow | tin | decomposition |
| | gold | none | heat |
| | pale yellow | none | decomposition |
| ***Mentzelia nuda*** BRACTLESS BLAZING-STAR | | | |
| flowers | light yellow | alum | heat |
| | antique gold | iron | heat |
| | bright golden yellow | tin | heat |
| ***Monarda fistulosa*** WILD BERGAMOT | | | |
| leaves | light yellow | alum | decomposition |
| | light yellow | tin | decomposition |
| | beige-yellow | none | heat |
| | pale yellow | none | decomposition |
| petals | pale yellow | alum | heat |
| | dull green-yellow | copper | heat |
| | green-yellow | tin | heat |
| | golden yellow | tin | decomposition |
| | pale yellow | none | heat |
| spikes | dull yellow | alum | heat |
| | pale yellow | alum | decomposition |
| | beige-yellow | copper | decomposition |
| | light yellow | tin | decomposition |
| | pale yellow | none | heat |
| ***Morus rubra*** RED MULBERRY | | | |
| bark | pale yellow | alum | heat |
| | pale yellow | tin | heat |
| bark green | bright yellow | alum | heat |
| | light yellow | alum | decomposition |
| | dull yellow | iron | decomposition |
| | bright yellow | tin | heat |
| | pale yellow | none | heat |
| branches | pale yellow | tin | heat |
| branchlets | bright yellow | alum | heat |
| | yellow | copper | heat |
| | dull yellow | iron | decomposition |

| DYE SOURCE | DYE RESULT | MORDANT | PROCESSING |
|---|---|---|---|
| ***Morus rubra,*** branchlets, continued | | | |
| | bright golden yellow | tin | heat |
| | cream-yellow | tin | decomposition |
| leaves | pale yellow | alum | heat |
| | pale yellow | alum | decomposition |
| | light dull yellow | copper | heat |
| | dull yellow | iron | decomposition |
| | cream-yellow | tin | heat |
| | cream-yellow | tin | decomposition |
| ***Neptunia lutea*** YELLOW NEPTUNE | | | |
| flower buds | dull yellow | alum | heat |
| | light olive-gold | copper | heat |
| | bright golden yellow | tin | heat |
| | dull yellow | none | heat |
| flowers | dull yellow | alum | heat |
| | dull yellow | copper | heat |
| | bright golden yellow | tin | heat |
| | dull yellow | none | heat |
| leaves | yellow | alum | heat |
| | light yellow | alum | decomposition |
| | bright golden yellow | tin | heat |
| | bright golden yellow | tin | decomposition |
| ***Nuttallanthus canadensis*** TOAD-FLAX | | | |
| leaves and stems | pale yellow | copper | decomposition |
| | antique gold | tin | heat |
| petals | golden yellow | tin | heat |
| ***Nyssa sylvatica*** BLACK GUM | | | |
| bark | pale yellow | tin | heat |
| | pale yellow | tin | decomposition |
| branchlets | dull yellow | tin | decomposition |
| leaves | pale yellow | alum | heat |
| | yellow | tin | heat |
| | golden yellow | tin | decomposition |
| ***Oenothera heterophylla*** SAND EVENING-PRIMROSE | | | |
| flowers | dull yellow | alum | heat |
| | bright golden yellow | tin | heat |
| leaves | dull yellow | alum | heat |
| | bright golden yellow | tin | heat |
| stems | bright golden yellow | tin | heat |
| ***Oenothera macrocarpa*** BIGFRUIT EVENING-PRIMROSE | | | |
| bracts and stems | yellow | tin | heat |
| | light yellow | tin | decomposition |
| fruits mature | dull yellow | tin | heat |
| fruits immature | yellow | tin | heat |
| | light tan-yellow | tin | decomposition |
| leaves | bright yellow | tin | heat |
| | bright golden yellow | tin | decomposition |
| petals | pale yellow | alum | decomposition |
| | beige-yellow | copper | heat |
| | dull yellow | tin | heat |

| DYE SOURCE | DYE RESULT | MORDANT | PROCESSING |
|---|---|---|---|
| ***Opuntia macrorhiza*** PRICKLY-PEAR | | | |
| fruits | light golden yellow | tin | heat |
| | pale yellow | tin | decomposition |
| ***Packera obovata*** ROUNDLEAF GROUNDSEL | | | |
| flowers | pale yellow | alum | heat |
| | pale yellow | alum | decomposition |
| | bright yellow | tin | heat |
| | bright yellow | tin | decomposition |
| | pale yellow | none | heat |
| leaves and stems | dull yellow | alum | heat |
| | dull yellow | alum | decomposition |
| | dull yellow | copper | heat |
| | light olive-gold | copper | decomposition |
| | bright golden yellow | tin | heat |
| | bright golden yellow | tin | decomposition |
| | light dull yellow | none | heat |
| ***Parthenocissus quinquefolia*** VIRGINIA CREEPER | | | |
| bark | gold | tin | heat |
| fruits | pale yellow | alum | heat |
| | cream-yellow | alum | decomposition |
| | dull yellow | copper | decomposition |
| | pale yellow | tin | heat |
| | cream-yellow | tin | decomposition |
| | cream-yellow | none | decomposition |
| leaves | pale yellow | alum | decomposition |
| | dull yellow | copper | heat |
| | pale yellow | copper | decomposition |
| | pale yellow | iron | decomposition |
| | bright golden yellow | tin | heat |
| | light yellow | tin | decomposition |
| | pale yellow | none | decomposition |
| pedicels | pale yellow | alum | heat |
| | light dull yellow | copper | heat |
| | light yellow | tin | heat |
| ***Pediomelum cuspidatum*** TALL-BREAD SCURFPEA | | | |
| fruits | light olive-gold | iron | heat |
| | dull yellow | tin | heat |
| | cream-yellow | tin | decomposition |
| leaves | pale yellow | alum | heat |
| | dull yellow | copper | heat |
| | bright yellow | tin | heat |
| | pale yellow | none | heat |
| stems | pale yellow | alum | decomposition |
| | dull yellow | iron | heat |
| | dull yellow | tin | heat |
| | light yellow | tin | decomposition |
| | pale yellow | none | decomposition |
| ***Phlox pilosa*** PRAIRIE PHLOX | | | |
| leaves | light yellow | alum | heat |
| | dull yellow | copper | heat |

| DYE SOURCE | DYE RESULT | MORDANT | PROCESSING |
|---|---|---|---|
| ***Phlox pilosa,*** leaves, continued | | | |
| | light golden yellow | tin | heat |
| | yellow | tin | decomposition |
| | pale yellow | none | heat |
| petals | light yellow | alum | heat |
| | bright golden yellow | tin | heat |
| stems | pale yellow | alum | decomposition |
| | bright yellow | tin | heat |
| ***Phyla lanceolata*** LANCELEAF FROG-FRUIT | | | |
| flowers | light yellow | alum | heat |
| | dull yellow | copper | heat |
| | light yellow | tin | heat |
| leaves | light yellow | alum | heat |
| | light yellow | alum | decomposition |
| | dull yellow | copper | heat |
| | light yellow | tin | heat |
| | cream-yellow | tin | decomposition |
| | pale yellow | none | heat |
| | pale yellow | none | decomposition |
| stems | light yellow | alum | heat |
| | pale yellow | tin | heat |
| | pale yellow | none | heat |
| ***Physalis angulata*** CUTLEAF GROUND-CHERRY | | | |
| fruits | yellow | tin | heat |
| leaves | pale yellow | alum | heat |
| | light dull yellow | alum | decomposition |
| | light golden yellow | tin | heat |
| | light golden yellow | tin | decomposition |
| stems | dull yellow | tin | heat |
| | light yellow | tin | decomposition |
| ***Physalis virginiana*** VIRGINIA GROUND-CHERRY | | | |
| leaves | dull yellow | alum | heat |
| | light yellow | alum | decomposition |
| | beige-yellow | copper | heat |
| | light yellow | copper | decomposition |
| | dull yellow | iron | decomposition |
| | light golden yellow | tin | decomposition |
| | pale yellow | none | decomposition |
| stems | light yellow | tin | heat |
| ***Phytolacca americana*** POKEWEED | | | |
| leaves | pale yellow | alum | heat |
| | pale yellow | alum | decomposition |
| | pale yellow | copper | decomposition |
| | pale yellow | iron | decomposition |
| | light yellow | tin | heat |
| | pale yellow | tin | decomposition |
| | pale yellow | none | heat |
| | pale yellow | none | decomposition |
| roots | light yellow | tin | decomposition |
| stems | pale yellow | alum | decomposition |

| DYE SOURCE | DYE RESULT | MORDANT | PROCESSING |
|---|---|---|---|
| | pale yellow | iron | decomposition |
| | light yellow | tin | decomposition |
| ***Platanus occidentalis*** SYCAMORE | | | |
| bark of branches | golden yellow | tin | heat |
| ***Podophyllum peltatum*** MAY-APPLE | | | |
| leaves | dull yellow | copper | heat |
| | light olive-gold | iron | heat |
| | bright golden yellow | tin | heat |
| | yellow | tin | decomposition |
| rhizomes and roots | yellow | alum | heat |
| | light antique olive-gold | copper | heat |
| | bright golden yellow | tin | heat |
| | dull yellow | tin | decomposition |
| ***Polygala alba*** WHITE MILKWORT | | | |
| flowers | pale yellow | alum | decomposition |
| | dull yellow | tin | heat |
| | light yellow | tin | decomposition |
| leaves and stems | light yellow | alum | decomposition |
| | dull yellow | copper | decomposition |
| | light olive-gold | iron | decomposition |
| | dull yellow | tin | heat |
| | bright golden yellow | tin | decomposition |
| | pale yellow | none | decomposition |
| ***Polygonum pensylvanicum*** PENNSYLVANIA SMARTWEED | | | |
| flowers and flower buds | light green-yellow | alum | heat |
| | light yellow | alum | decomposition |
| | bright gold | tin | heat |
| | light golden yellow | tin | decomposition |
| leaves | bright yellow | tin | heat |
| stems | light dull yellow | alum | heat |
| | yellow | tin | heat |
| | light golden yellow | tin | decomposition |
| ***Polygonum punctatum*** DOTTED SMARTWEED | | | |
| flowers and flower buds | green-yellow | alum | heat |
| | bright golden yellow | tin | heat |
| leaves | yellow | alum | heat |
| | dull yellow | alum | decomposition |
| | light dull yellow | copper | decomposition |
| | bright golden yellow | tin | heat |
| | bright yellow | tin | decomposition |
| stems | pale yellow | alum | heat |
| | pale yellow | alum | decomposition |
| | light golden yellow | tin | heat |
| ***Polystichum acrostichoides*** CHRISTMAS FERN | | | |
| leaves | yellow | tin | heat |
| | pale yellow | tin | decomposition |
| ***Polytaenia nuttallii*** PRAIRIE PARSLEY | | | |
| flowers | pale yellow | alum | heat |
| | cream-yellow | alum | decomposition |
| | light olive-gold | iron | heat |

| DYE SOURCE | DYE RESULT | MORDANT | PROCESSING |
|---|---|---|---|
| ***Polytaenia nuttallii,*** flowers, continued | | | |
| | bright golden yellow | tin | heat |
| | light golden yellow | tin | decomposition |
| | pale yellow | none | heat |
| leaves and stems | dull yellow | alum | heat |
| | pale yellow | alum | decomposition |
| | dull yellow | copper | heat |
| | pale yellow | copper | decomposition |
| | antique gold | iron | heat |
| | bright golden yellow | tin | heat |
| | light golden yellow | tin | decomposition |
| | dull yellow | none | heat |
| | pale yellow | none | decomposition |
| roots | dull yellow | iron | heat |
| | dull yellow | iron | decomposition |
| | pale yellow | tin | decomposition |
| ***Populus deltoides*** EASTERN COTTONWOOD | | | |
| leaves | light yellow | alum | heat |
| | pale yellow | alum | decomposition |
| | dull yellow | copper | heat |
| | bright yellow | tin | heat |
| | light yellow | tin | decomposition |
| | pale yellow | none | heat |
| stems | yellow | alum | heat |
| | antique gold | copper | heat |
| | cream-yellow | tin | decomposition |
| | dull yellow | none | heat |
| ***Potentilla arguta*** TALL CINQUEFOIL | | | |
| flowers | gold | alum | heat |
| | dull yellow | alum | decomposition |
| | antique gold | copper | heat |
| | dull yellow | copper | decomposition |
| | yellow | tin | heat |
| | light tan-yellow | tin | decomposition |
| leaves | dull yellow | alum | heat |
| | light golden yellow | tin | heat |
| stems | dull yellow | alum | heat |
| | yellow | tin | heat |
| ***Prosopis glandulosa*** MESQUITE | | | |
| roots | gold | tin | heat |
| | cream-yellow | tin | decomposition |
| ***Psoralidium tenuiflorum*** SLIM-FLOWER SCURFPEA | | | |
| flowers | light olive-gold | iron | heat |
| leaves | light yellow | alum | heat |
| | bright yellow | alum | decomposition |
| | dull yellow | copper | heat |
| | bright yellow | tin | heat |
| | dull yellow | tin | decomposition |
| | pale yellow | none | heat |

| DYE SOURCE | DYE RESULT | MORDANT | PROCESSING |
|---|---|---|---|
| ***Ptilimnium nuttallii*** MOCK BISHOP'S-WEED | | | |
| leaves and stems | yellow | alum | heat |
| | light yellow | alum | decomposition |
| | dull yellow | copper | decomposition |
| | dull yellow | iron | decomposition |
| | bright golden yellow | tin | heat |
| | light golden yellow | tin | decomposition |
| | dull yellow | none | heat |
| | pale yellow | none | decomposition |
| roots | dull yellow | alum | heat |
| | bright yellow | tin | heat |
| ***Pyrrhopappus grandiflorus*** MORNING STAR | | | |
| leaves | pale yellow | alum | heat |
| | pale yellow | alum | decomposition |
| | light olive | copper | heat |
| | cream-yellow | copper | decomposition |
| | beige-yellow | iron | decomposition |
| | bright yellow | tin | heat |
| | bright yellow | tin | decomposition |
| ligulate florets | bright yellow | alum | heat |
| | yellow | alum | decomposition |
| | bright gold | copper | heat |
| | yellow | copper | decomposition |
| | antique gold | iron | decomposition |
| | yellow | none | heat |
| | yellow | none | decomposition |
| ***Ratibida columnifera*** MEXICAN HAT | | | |
| disk florets | antique olive-gold | tin | heat |
| leaves and stems | bright yellow | alum | heat |
| | dull yellow | copper | heat |
| | light golden yellow | tin | heat |
| | cream-yellow | tin | decomposition |
| ray florets | yellow | tin | heat |
| roots | light olive-gold | iron | heat |
| ***Rhus copallinum*** WINGED SUMAC | | | |
| flowers | yellow | alum | heat |
| | bright golden yellow | tin | heat |
| | dull yellow | none | decomposition |
| leaves | dull yellow | alum | heat |
| | dull yellow | alum | decomposition |
| | light olive-gold | copper | decomposition |
| | bright yellow | tin | heat |
| | light tan-yellow | tin | decomposition |
| ***Rhus glabra*** SMOOTH SUMAC | | | |
| bark | bright golden yellow | tin | heat |
| fruits | antique gold | copper | heat |
| leaves | dull yellow | alum | heat |
| | light yellow | alum | decomposition |
| | dull yellow | copper | decomposition |

| DYE SOURCE | DYE RESULT | MORDANT | PROCESSING |
|---|---|---|---|
| ***Rhus glabra,*** leaves, continued | | | |
| | light golden yellow | tin | heat |
| | light tan-yellow | tin | decomposition |
| | light dull yellow | none | heat |
| ***Robinia hispida*** BRISTLY LOCUST | | | |
| flowers | pale yellow | alum | decomposition |
| | bright golden yellow | tin | heat |
| | bright yellow | tin | decomposition |
| leaves | bright yellow | alum | heat |
| | pale yellow | alum | decomposition |
| | beige-yellow | copper | decomposition |
| | dull yellow | tin | decomposition |
| | beige-yellow | none | heat |
| stems | bright golden yellow | tin | heat |
| ***Robinia pseudoacacia*** BLACK LOCUST | | | |
| bark | pale yellow | tin | decomposition |
| leaves | dull yellow | tin | heat |
| | light dull yellow | none | heat |
| petals | pale yellow | alum | heat |
| | light bright green-yellow | tin | heat |
| | pale yellow | tin | decomposition |
| ***Rudbeckia grandiflora*** ROUGH CONEFLOWER | | | |
| disk florets | dull yellow | alum | heat |
| | golden yellow | tin | heat |
| | dull yellow | tin | decomposition |
| leaves and stems | beige-yellow | copper | heat |
| ray florets | yellow | tin | heat |
| | light yellow | tin | decomposition |
| rhizomes | pale yellow | tin | heat |
| | pale yellow | tin | decomposition |
| ***Rudbeckia hirta*** BLACK-EYED SUSAN | | | |
| leaves | bright yellow | alum | heat |
| | light antique olive-gold | copper | heat |
| | bright gold | tin | heat |
| | pale yellow | tin | decomposition |
| | dull yellow | none | heat |
| ray florets | light yellow | alum | heat |
| | dull yellow | copper | heat |
| | antique gold | iron | heat |
| | bright golden yellow | tin | heat |
| | bright yellow | tin | decomposition |
| roots | dull yellow | tin | heat |
| | pale yellow | tin | decomposition |
| ***Rumex altissimus*** PALE DOCK | | | |
| flowers | bright gold | alum | heat |
| | antique gold | copper | heat |
| | gold | none | heat |
| leaves | dull yellow | copper | heat |
| | bright yellow | tin | decomposition |
| stems | antique olive-gold | alum | heat |
| | light olive-gold | none | heat |

| DYE SOURCE | DYE RESULT | MORDANT | PROCESSING |
|---|---|---|---|
| ***Sabatia campestris*** PRAIRIE ROSE GENTIAN | | | |
| leaves and stems | dull yellow | alum | decomposition |
| | cream-yellow | copper | decomposition |
| | bright yellow | tin | heat |
| | bright yellow | tin | decomposition |
| | light yellow | none | decomposition |
| petals | pale yellow | alum | heat |
| | cream-yellow | alum | decomposition |
| | dull yellow | tin | decomposition |
| ***Salix caroliniana*** CAROLINA WILLOW | | | |
| leaves | light yellow | alum | heat |
| | light yellow | alum | decomposition |
| | dull yellow | copper | heat |
| | pale yellow | copper | decomposition |
| | bright golden yellow | tin | heat |
| | bright golden yellow | tin | decomposition |
| ***Salix exigua*** SANDBAR WILLOW | | | |
| leaves | light bright green-yellow | alum | heat |
| | light yellow | alum | decomposition |
| | dull yellow | copper | decomposition |
| | dull yellow | iron | decomposition |
| | light golden yellow | tin | heat |
| | bright golden yellow | tin | decomposition |
| | pale yellow | none | heat |
| | pale yellow | none | decomposition |
| stems | light tan-gold | tin | heat |
| ***Salix nigra*** BLACK WILLOW | | | |
| branchlets | light tan-yellow | alum | heat |
| | dull yellow | copper | decomposition |
| | cream-yellow | iron | decomposition |
| | bright golden yellow | tin | heat |
| leaves | bright yellow | alum | heat |
| | pale yellow | alum | decomposition |
| | dull yellow | copper | heat |
| | bright yellow | tin | heat |
| | bright yellow | tin | decomposition |
| ***Sambucus canadensis*** ELDERBERRY | | | |
| bark | golden yellow | tin | heat |
| leaves | dull yellow | alum | heat |
| | light olive-gold | copper | heat |
| | bright antique gold | tin | heat |
| | dull yellow | tin | decomposition |
| ***Sanguinaria canadensis*** BLOODROOT | | | |
| leaves and stems | light olive-gold | iron | heat |
| | golden yellow | tin | heat |
| | dull yellow | tin | decomposition |
| ***Sapindus drummondii*** SOAPBERRY | | | |
| bark | pale yellow | alum | decomposition |
| | pale yellow | tin | decomposition |
| flowers | bright yellow | alum | heat |

| DYE SOURCE | DYE RESULT | MORDANT | PROCESSING |
|---|---|---|---|
| ***Sapindus drummondii,*** flowers, continued | | | |
| | light yellow | alum | decomposition |
| | dull yellow | copper | heat |
| | light olive-gold | copper | decomposition |
| | bright golden yellow | tin | decomposition |
| | pale yellow | none | decomposition |
| fruits | light yellow | alum | decomposition |
| | pale yellow | copper | heat |
| | dull yellow | copper | decomposition |
| | dull yellow | iron | decomposition |
| | cream-yellow | tin | heat |
| | bright golden yellow | tin | decomposition |
| | pale yellow | none | decomposition |
| leaves | green-yellow | alum | heat |
| | pale yellow | alum | decomposition |
| | bright gold | tin | heat |
| | dull yellow | tin | decomposition |
| stems | dull yellow | alum | heat |
| | pale yellow | alum | decomposition |
| | dull yellow | copper | heat |
| | gold | iron | heat |
| | bright golden yellow | tin | heat |
| | light yellow | tin | decomposition |
| ***Sassafras albidum*** SASSAFRAS | | | |
| bark | light tan-gold | alum | decomposition |
| | light tan-gold | copper | decomposition |
| | antique gold | iron | decomposition |
| | bright orange-gold | tin | decomposition |
| flowers | dull yellow | alum | heat |
| | light yellow | alum | decomposition |
| | light olive-gold | copper | heat |
| | dull yellow | copper | decomposition |
| | bright golden yellow | tin | heat |
| | bright golden yellow | tin | decomposition |
| | pale yellow | none | heat |
| leaves | light yellow | alum | heat |
| | pale yellow | alum | decomposition |
| | dull yellow | copper | decomposition |
| | dull yellow | iron | decomposition |
| | golden yellow | tin | heat |
| | bright golden yellow | tin | decomposition |
| | pale yellow | none | heat |
| roots | beige-yellow | copper | decomposition |
| | light yellow | tin | decomposition |
| ***Sedum nuttallianum*** YELLOW STONECROP | | | |
| fruits, flowers, and stems | bright gold | tin | heat |
| ***Senecio riddellii*** SAND GROUNDSEL | | | |
| flowers | light antique olive-gold | tin | heat |
| | gold | tin | decomposition |
| leaves | light antique olive-gold | tin | heat |
| | bright yellow | tin | decomposition |

| DYE SOURCE | DYE RESULT | MORDANT | PROCESSING |
|---|---|---|---|
| ***Silene stellata*** STARRY CAMPION | | | |
| flowers without sepals | cream-yellow | tin | heat |
| leaves and upper stems | cream-yellow | tin | heat |
| stems (basal) and rootstocks | light yellow | tin | heat |
| ***Silphium integrifolium*** PRAIRIE ROSINWEED | | | |
| disk and ray florets | yellow | tin | heat |
| | pale yellow | tin | decomposition |
| | pale yellow | none | decomposition |
| leaves | bright golden yellow | alum | heat |
| | antique olive-gold | copper | heat |
| | bright golden yellow | tin | heat |
| | dull yellow | none | heat |
| roots | light olive-gold | iron | decomposition |
| | yellow | tin | heat |
| | dull yellow | tin | decomposition |
| ***Sisyrinchium angustifolium*** BLUE-EYED GRASS | | | |
| leaves | dull yellow | copper | heat |
| | antique gold | iron | decomposition |
| | bright antique gold | tin | decomposition |
| | beige-yellow | none | decomposition |
| ***Smilax bona-nox*** GREENBRIER | | | |
| leaves and stems | pale yellow | alum | decomposition |
| | bright golden yellow | tin | heat |
| | yellow | tin | decomposition |
| ***Solanum dimidiatum*** WESTERN HORSE-NETTLE | | | |
| leaves and stems | green-yellow | alum | heat |
| petals | beige-yellow | iron | decomposition |
| | light yellow | tin | heat |
| | light yellow | tin | decomposition |
| roots | dull yellow | tin | heat |
| | light yellow | tin | decomposition |
| stamen | light yellow | tin | heat |
| ***Solanum elaeagnifolium*** SILVERLEAF NIGHTSHADE | | | |
| leaves | dull yellow | alum | heat |
| | bright golden yellow | tin | heat |
| | pale yellow | none | heat |
| petals | light dull yellow | copper | heat |
| | yellow | tin | heat |
| ***Solidago missouriensis*** PLAINS GOLDENROD | | | |
| heads | bright yellow | alum | heat |
| | light bright green-yellow | alum | decomposition |
| | yellow | copper | decomposition |
| | bright golden yellow | tin | decomposition |
| | dull yellow | none | heat |
| | light yellow | none | decomposition |
| leaves | bright yellow | alum | heat |
| | light yellow | alum | decomposition |
| | light green-yellow | copper | decomposition |
| | bright antique gold | tin | heat |
| | bright golden yellow | tin | decomposition |

| DYE SOURCE | DYE RESULT | MORDANT | PROCESSING |
|---|---|---|---|
| ***Solidago missouriensis,*** leaves, continued | | | |
| | light yellow | none | decomposition |
| rhizomes and roots | bright yellow | tin | heat |
| | pale yellow | tin | decomposition |
| stems | yellow | alum | heat |
| | bright yellow | alum | decomposition |
| | bright golden yellow | tin | heat |
| | bright golden yellow | tin | decomposition |
| ***Solidago rigida*** STIFF GOLDENROD | | | |
| heads immature | light bright green-yellow | alum | heat |
| | yellow | alum | decomposition |
| | dull yellow | copper | decomposition |
| | bright yellow | tin | heat |
| | bright yellow | tin | decomposition |
| | light dull yellow | none | heat |
| | dull yellow | none | decomposition |
| heads mature | light bright green-yellow | alum | heat |
| | light yellow | alum | decomposition |
| | dull yellow | copper | decomposition |
| | bright golden yellow | tin | heat |
| | bright yellow | tin | decomposition |
| leaves | bright yellow | alum | heat |
| | light yellow | alum | decomposition |
| | dull yellow | copper | decomposition |
| | bright yellow | tin | heat |
| | bright yellow | tin | decomposition |
| | pale yellow | none | heat |
| | pale yellow | none | decomposition |
| roots | beige-yellow | iron | decomposition |
| | light yellow | tin | heat |
| stems | bright yellow | alum | heat |
| | bright yellow | tin | heat |
| | light yellow | tin | decomposition |
| | pale yellow | none | heat |
| ***Sorghastrum nutans*** INDIANGRASS | | | |
| leaves | light yellow | alum | heat |
| | bright golden yellow | tin | heat |
| | pale yellow | tin | decomposition |
| rhizomes and roots | cream-yellow | tin | heat |
| spikelets | dull yellow | alum | heat |
| | pale yellow | alum | decomposition |
| | dull yellow | iron | heat |
| | bright gold | tin | heat |
| | pale yellow | tin | decomposition |
| stems | pale yellow | copper | heat |
| | cream-yellow | tin | heat |
| ***Stillingia sylvatica*** QUEEN'S DELIGHT | | | |
| bark green | bright green-yellow | copper | heat |
| leaves | light antique olive-gold | alum | heat |
| | dull yellow | alum | decomposition |
| | bright golden yellow | tin | heat |

| DYE SOURCE | DYE RESULT | MORDANT | PROCESSING |
|---|---|---|---|
| | bright gold | tin | decomposition |
| | dull yellow | none | decomposition |
| pistillate flowers | light antique olive-gold | alum | heat |
| | golden yellow | tin | heat |
| roots | gold | alum | heat |
| | pale yellow | alum | decomposition |
| | bright golden yellow | tin | heat |
| | pale yellow | tin | decomposition |
| staminate flowers | light olive-gold | alum | heat |
| | light yellow | alum | decomposition |
| | golden yellow | tin | decomposition |
| | dull yellow | none | decomposition |
| stems | dull yellow | alum | heat |
| | light olive-gold | copper | heat |
| | bright golden yellow | tin | heat |
| | light golden yellow | tin | decomposition |
| ***Streptanthus hyacinthoides*** SMOOTH TWIST-FLOWER | | | |
| flowers | dull yellow | tin | heat |
| ***Tephrosia virginiana*** GOAT'S RUE | | | |
| leaves | light yellow | alum | heat |
| | pale yellow | alum | decomposition |
| | yellow | copper | heat |
| | light golden yellow | tin | heat |
| | light yellow | tin | decomposition |
| rhizomes | cream-yellow | tin | heat |
| ***Teucrium canadense*** AMERICAN GERMANDER | | | |
| flowers | beige-yellow | copper | heat |
| | light yellow | tin | heat |
| leaves | light green-yellow | alum | heat |
| | bright yellow | tin | heat |
| roots | beige-yellow | iron | heat |
| | dull yellow | tin | heat |
| ***Thelesperma filifolium*** GREENTHREAD | | | |
| leaves | dull yellow | alum | decomposition |
| | gold | copper | heat |
| | yellow | none | heat |
| | dull yellow | none | decomposition |
| ray florets | light golden yellow | alum | heat |
| | golden yellow | alum | decomposition |
| | golden yellow | copper | heat |
| | golden yellow | copper | decomposition |
| | light golden yellow | none | heat |
| | gold | none | decomposition |
| ***Tradescantia occidentalis*** PRAIRIE SPIDERWORT | | | |
| stems | pale yellow | alum | decomposition |
| | pale yellow | copper | decomposition |
| | gold | iron | heat |
| | yellow | tin | heat |
| | light yellow | tin | decomposition |
| | pale yellow | none | decomposition |

| DYE SOURCE | DYE RESULT | MORDANT | PROCESSING |
|---|---|---|---|
| ***Ulmus rubra*** SLIPPERY ELM | | | |
| bark | light golden yellow | tin | heat |
| leaves | bright golden yellow | tin | heat |
| | gold | tin | decomposition |
| ***Vaccinium arboreum*** FARKLEBERRY | | | |
| leaves | bright golden yellow | tin | decomposition |
| ***Vernonia baldwinii*** WESTERN IRONWEED | | | |
| leaves | yellow | alum | heat |
| | pale yellow | alum | decomposition |
| | pale yellow | copper | decomposition |
| | bright gold | tin | heat |
| | light yellow | tin | decomposition |
| | dull yellow | none | heat |
| | pale yellow | none | decomposition |
| stems | dull yellow | alum | heat |
| | yellow | tin | heat |
| ***Viburnum rufidulum*** RUSTY BLACK HAW | | | |
| flowers | pale yellow | alum | heat |
| | beige-yellow | copper | heat |
| | light olive-gold | iron | heat |
| | bright golden yellow | tin | heat |
| | bright yellow | tin | decomposition |
| leaves | bright golden yellow | tin | heat |
| | bright yellow | tin | decomposition |
| pedicels | cream-yellow | alum | decomposition |
| | dull yellow | iron | heat |
| | bright golden yellow | tin | heat |
| | bright golden yellow | tin | decomposition |
| ***Vitis aestivalis*** PIGEON GRAPE | | | |
| bark | pale yellow | tin | decomposition |
| leaves | pale yellow | alum | heat |
| | beige-yellow | iron | heat |
| | bright yellow | tin | heat |
| | cream-yellow | tin | decomposition |

# 8 ORANGE

## Nature's Brightest Dye Color

Our search for natural dyes resulted in various orange colors, shown here with Munsell notations. Following these color chips is a list of the plants that were used to obtain these colors, including mordant and processing conditions.

2.5YR 6/14

2.5YR 6/16

5YR 6/14

5YR 7/14

**BRIGHT ORANGE**

7.5YR 7/12

7.5YR 7/14

7.5YR 7/16

**BRIGHT YELLOW·ORANGE**

2.5YR 4/10

**BURNT ORANGE**

2.5YR 5/12

**DARK ORANGE**

5YR 3/6

5YR 4/8

**DARK TERRA·COTTA**

7.5YR 7/10

**DULL MEDIUM ORANGE**

2.5YR 7/6

5YR 7/6

**GRAYED SALMON·ORANGE**

6.25YR 6/12

**LIGHT BRIGHT ORANGE**

7.5YR 8/8

**LIGHT DULL ORANGE**

7.5YR 8/6

**LIGHT SALMON·ORANGE**

2.5YR 5/8 5YR 6/8 5YR 7/8

**LIGHT TERRA·COTTA**

7.5YR 8/10

**LIGHT YELLOW·ORANGE**

2.5YR 5/14 5YR 6/12 5YR 7/12

**ORANGE**

10YR 9/4

**PALE ORANGE**

1.25YR 5/12

**RED·ORANGE**

2.5YR 8/6 5YR 8/6

**SALMON·ORANGE**

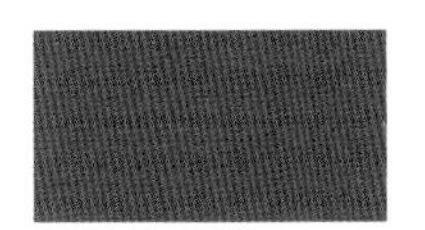 

1.25YR 4/12 3.75YR 5/12

**TERRA·COTTA**

| DYE SOURCE | DYE RESULT | MORDANT | PROCESSING |
|---|---|---|---|
| ***Acer negundo*** BOXELDER | | | |
| bark | pale orange | none | heat |
| ***Alnus maritima*** SEASIDE ALDER | | | |
| stems | pale orange | alum | decomposition |
| ***Apocynum cannabinum*** INDIAN HEMP | | | |
| roots | light salmon-orange | tin | heat |
| ***Argemone polyanthemos*** PRICKLY POPPY | | | |
| leaves | bright yellow-orange | tin | heat |
| ***Bidens aristosa*** BEARDED BEGGAR-TICKS | | | |
| disk and ray florets | burnt orange | tin | heat |
| leaves | light bright orange | tin | heat |
| ***Bidens bipinnata*** SPANISH NEEDLES | | | |
| leaves | bright yellow-orange | tin | heat |
| ***Calylophus hartwegii*** HARTWEG'S SUNDROP | | | |
| leaves | bright yellow-orange | tin | heat |
| ***Calylophus serrulatus*** HALFSHRUB SUNDROP | | | |
| roots | salmon-orange | tin | heat |
| ***Carya cordiformis*** BITTERNUT HICKORY | | | |
| branches | light salmon-orange | copper | heat |
| | pale orange | none | heat |
| fruits | grayed salmon-orange | tin | decomposition |
| ***Ceanothus americanus*** NEW JERSEY TEA | | | |
| rhizomes and roots | light terra-cotta | tin | heat |
| ***Cercis canadensis*** REDBUD | | | |
| bark | pale orange | tin | decomposition |
| | pale orange | none | decomposition |
| bark of branches | light salmon-orange | alum | heat |
| flowers | orange | tin | decomposition |
| ***Chrysopsis pilosa*** SOFT GOLDEN-ASTER | | | |
| disk and ray florets | bright orange | tin | heat |
| ***Cicuta maculata*** WATER-HEMLOCK | | | |
| flowers | bright yellow-orange | tin | heat |
| leaves | bright yellow-orange | tin | heat |
| ***Comandra umbellata*** BASTARD TOAD-FLAX | | | |
| leaves | bright yellow-orange | tin | heat |
| stems | bright yellow-orange | tin | heat |
| ***Coreopsis tinctoria*** PLAINS TICKSEED | | | |
| disk florets and bracts | dark terra-cotta | tin | heat |
| leaves | bright orange | tin | heat |
| ray florets | dark terra-cotta | tin | heat |
| ***Cornus drummondii*** ROUGH-LEAF DOGWOOD | | | |
| branchlets | pale orange | tin | heat |
| ***Cornus florida*** FLOWERING DOGWOOD | | | |
| bark | pale orange | alum | decomposition |
| | light salmon-orange | tin | decomposition |
| | pale orange | none | decomposition |
| leaves | light salmon-orange | alum | decomposition |
| | dark terra-cotta | tin | decomposition |

| DYE SOURCE | DYE RESULT | MORDANT | PROCESSING |
|---|---|---|---|
| ***Cotinus obovatus*** SMOKE-TREE | | | |
| bark | bright orange | tin | heat |
| wood | terra-cotta | alum | heat |
| | red-orange | tin | heat |
| ***Cyperus squarrosus*** BEARDED FLATSEDGE | | | |
| leaves | light yellow-orange | tin | heat |
| ***Diospyros virginiana*** PERSIMMON | | | |
| branchlets | bright yellow-orange | tin | heat |
| ***Hedyotis nigricans*** PRAIRIE BLUETS | | | |
| leaves and stems | orange | tin | heat |
| roots | pale orange | alum | heat |
| ***Helianthus tuberosus*** JERUSALEM ARTICHOKE | | | |
| disk and ray florets | bright yellow-orange | tin | heat |
| ***Ipomopsis rubra*** STANDING CYPRESS | | | |
| leaves | bright yellow-orange | tin | heat |
| ***Juglans nigra*** BLACK WALNUT | | | |
| fruits | pale orange | none | heat |
| involucral husks | light terra-cotta | tin | decomposition |
| leaves | bright yellow-orange | tin | heat |
| ***Liquidambar styraciflua*** SWEET GUM | | | |
| bark | light terra-cotta | alum | decomposition |
| leaves | bright yellow-orange | tin | heat |
| | dark orange | tin | decomposition |
| roots | grayed salmon-orange | tin | decomposition |
| ***Nyssa sylvatica*** BLACK GUM | | | |
| leaves | salmon-orange | none | decomposition |
| ***Oenothera heterophylla*** SAND EVENING-PRIMROSE | | | |
| flowers | light dull orange | alum | decomposition |
| | dull medium orange | tin | decomposition |
| leaves | orange | tin | decomposition |
| roots | grayed salmon-orange | alum | decomposition |
| stems | light terra-cotta | tin | decomposition |
| ***Parthenocissus quinquefolia*** VIRGINIA CREEPER | | | |
| leaves | salmon-orange | alum | heat |
| ***Physalis virginiana*** VIRGINIA GROUND-CHERRY | | | |
| leaves | bright yellow-orange | tin | heat |
| ***Populus deltoides*** EASTERN COTTONWOOD | | | |
| roots | pale orange | alum | heat |
| stems | bright orange | tin | heat |
| ***Potentilla arguta*** TALL CINQUEFOIL | | | |
| rootstocks | light salmon-orange | alum | heat |
| | light dull orange | tin | heat |
| | pale orange | tin | decomposition |
| ***Pyrrhopappus grandiflorus*** MORNING STAR | | | |
| ligulate florets | bright orange | tin | heat |
| | bright orange | tin | decomposition |
| ***Robinia hispida*** BRISTLY LOCUST | | | |
| leaves | bright yellow-orange | tin | heat |

| DYE SOURCE | DYE RESULT | MORDANT | PROCESSING |
|---|---|---|---|
| ***Rumex altissimus*** PALE DOCK | | | |
| flowers | bright orange | tin | heat |
| leaves | bright yellow-orange | tin | heat |
| stems | bright yellow-orange | tin | heat |
| ***Sabatia campestris*** PRAIRIE ROSE GENTIAN | | | |
| petals | salmon-orange | none | decomposition |
| ***Salix caroliniana*** CAROLINA WILLOW | | | |
| branches | light dull orange | tin | heat |
| ***Salix nigra*** BLACK WILLOW | | | |
| bark | pale orange | tin | heat |
| ***Sanguinaria canadensis*** BLOODROOT | | | |
| rhizomes and roots | terra-cotta | tin | heat |
| | bright orange | tin | decomposition |
| | dull medium orange | none | heat |
| | light dull orange | none | decomposition |
| ***Sapindus drummondii*** SOAPBERRY | | | |
| flowers | bright yellow-orange | tin | heat |
| ***Sassafras albidum*** SASSAFRAS | | | |
| bark | bright yellow-orange | tin | heat |
| ***Stillingia sylvatica*** QUEEN'S DELIGHT | | | |
| staminate flowers | bright yellow-orange | tin | heat |
| ***Thelesperma filifolium*** GREENTHREAD | | | |
| leaves | bright orange | tin | heat |
| | bright orange | tin | decomposition |
| ray florets | orange | tin | heat |
| ***Vaccinium arboreum*** FARKLEBERRY | | | |
| rootstocks | pale orange | tin | heat |
| ***Viburnum rufidulum*** RUSTY BLACK HAW | | | |
| bark | dull medium orange | tin | heat |
| ***Vitis aestivalis*** PIGEON GRAPE | | | |
| bark | light salmon-orange | tin | heat |

# 9 BROWN

## Nature's Autumn Dye Color

Our search for natural dyes resulted in various brown colors, shown here with Munsell notations. Following these color chips is a list of the plants that were used to obtain these colors, including mordant and processing conditions.

7.5YR 2/2 7.5Y 2/2 5Y 2/2

**BLACK·BROWN**

10YR 6/10 10YR 6/12

**BRIGHT GOLDEN TAN**

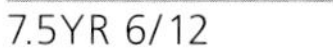

7.5YR 6/12 7.5YR 6/14

**BRIGHT ORANGE·TAN**

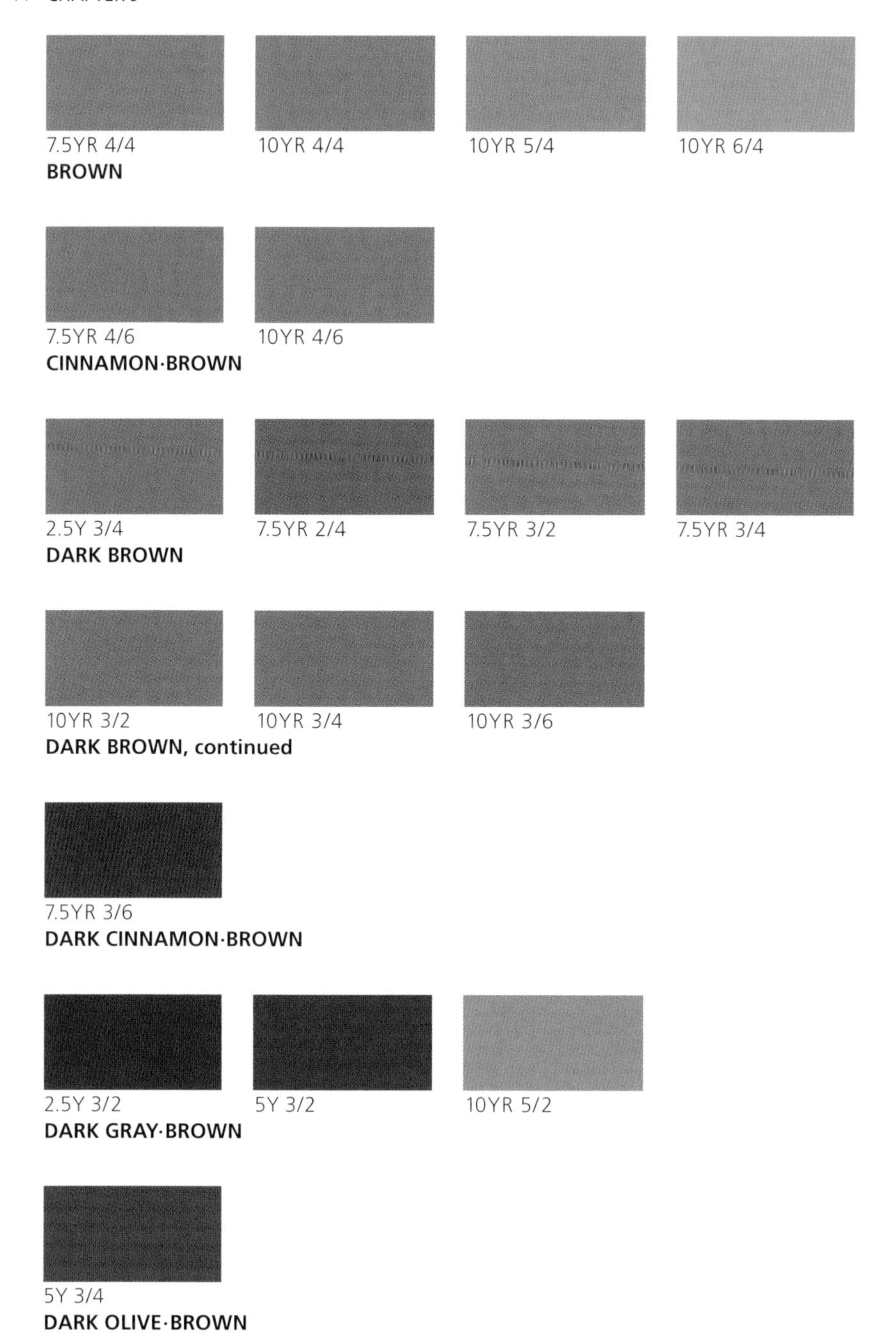
7.5YR 4/4
10YR 4/4
10YR 5/4
10YR 6/4
BROWN
7.5YR 4/6
10YR 4/6
CINNAMON·BROWN
2.5Y 3/4
7.5YR 2/4
7.5YR 3/2
7.5YR 3/4
DARK BROWN
10YR 3/2
10YR 3/4
10YR 3/6
DARK BROWN, continued
7.5YR 3/6
DARK CINNAMON·BROWN
2.5Y 3/2
5Y 3/2
10YR 5/2
DARK GRAY·BROWN
5Y 3/4
DARK OLIVE·BROWN

7.5YR 5/2

**DARK PINK GRAY·BROWN**

5YR 3/4

**DARK RED·BROWN**

10YR 5/6

10YR 5/8

**GOLDEN BROWN**

2.5Y 7/6

10YR 6/8

**GOLDEN TAN**

2.5Y 4/6

2.5Y 6/4

**LIGHT BROWN**

7.5YR 5/6

**LIGHT CINNAMON·BROWN**

2.5Y 6/6

**LIGHT GOLDEN BROWN**

2.5Y 8/4

10YR 7/8

**LIGHT GOLDEN TAN**

5Y 7/4

**LIGHT KHAKI**

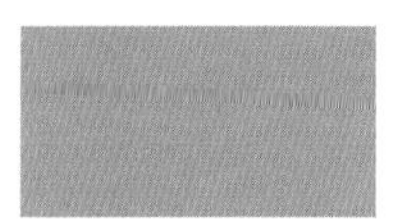
7.5YR 6/6

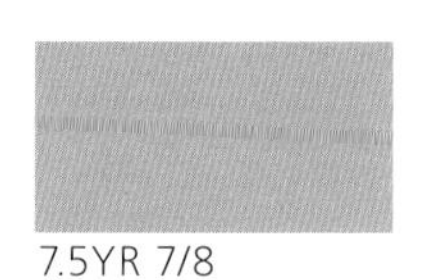
7.5YR 7/8

**LIGHT ORANGE·TAN**

7.5YR 7/4

**LIGHT PINK·BROWN**

7.5YR 7/6

**LIGHT SALMON·TAN**

2.5Y 8.5/4

10YR 7/4

**LIGHT TAN**

10YR 8/4

**LIGHT TAN·PEACH**

2.5Y 8.5/6 10YR 8/6

**LIGHT YELLOW·TAN**

 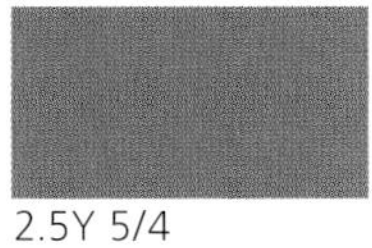 

2.5Y 4/4 2.5Y 5/4 7.5YR 5/4

**MEDIUM BROWN**

2.5Y 4/2 2.5Y 5/2

**MEDIUM GRAY·BROWN**

10YR 6/2

**MEDIUM GRAY·TAN**

7.5YR 6/2

**MEDIUM PINK·GRAY·BROWN**

7.5YR 6/4

**MEDIUM PINK·BROWN**

5YR 7/4

**MEDIUM PINK·TAN**

5Y 4/4

**OLIVE·BROWN**

7.5YR 5/8

7.5YR 5/10

**ORANGE·BROWN**

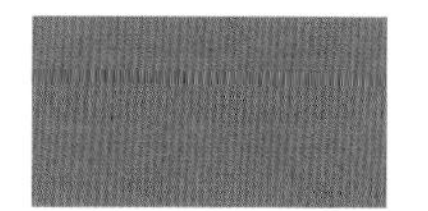

5YR 6/6

7.5YR 6/8

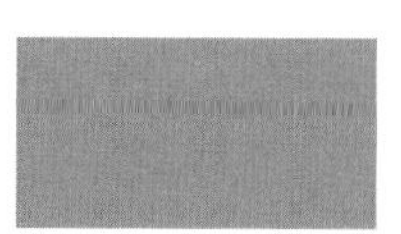

7.5YR 6/10

**ORANGE·TAN**

2.5YR 6/6

5YR 5/8

**PINK·BROWN**

2.5Y 7/4

**TAN**

2.5Y 8/6

10YR 6/6

10YR 7/6

**YELLOW·TAN**

| DYE SOURCE | DYE RESULT | MORDANT | PROCESSING |
|---|---|---|---|
| ***Acer negundo*** BOXELDER | | | |
| bark | light yellow-tan | alum | heat |
| | tan | copper | heat |
| ***Acer saccharinum*** SILVER MAPLE | | | |
| bark | light tan | alum | heat |
| | tan | copper | heat |

| DYE SOURCE | DYE RESULT | MORDANT | PROCESSING |
|---|---|---|---|
| | light yellow-tan | tin | heat |
| | light golden tan | none | decomposition |
| leaves | light golden tan | alum | decomposition |
| | light golden brown | copper | decomposition |
| | dark brown | iron | heat |
| | light golden tan | none | heat |
| | light golden tan | none | decomposition |
| ***Ageratina altissima*** WHITE SNAKEROOT | | | |
| leaves | tan | copper | decomposition |
| | light khaki | iron | decomposition |
| ***Alnus maritima*** SEASIDE ALDER | | | |
| leaves | olive-brown | iron | heat |
| stems | light tan | alum | heat |
| | tan | copper | heat |
| | light golden tan | copper | decomposition |
| | light tan-peach | none | decomposition |
| ***Ambrosia psilostachya*** WESTERN RAGWEED | | | |
| leaves and heads | tan | copper | decomposition |
| | olive-brown | iron | heat |
| stems | light golden tan | alum | decomposition |
| | tan | copper | decomposition |
| | light golden tan | none | decomposition |
| ***Ambrosia trifida*** GIANT RAGWEED | | | |
| leaves | medium brown | iron | heat |
| ***Amorpha canescens*** LEADPLANT | | | |
| bark | yellow-tan | alum | heat |
| | light golden tan | copper | heat |
| | golden tan | iron | heat |
| leaves | medium brown | iron | heat |
| roots | light golden tan | alum | heat |
| | tan | copper | heat |
| | light golden brown | iron | heat |
| | light tan | none | heat |
| spikes | tan | copper | heat |
| | light khaki | copper | decomposition |
| ***Apocynum cannabinum*** INDIAN HEMP | | | |
| leaves | olive-brown | iron | heat |
| roots | light tan-peach | copper | heat |
| | tan | iron | heat |
| stems | light tan-peach | copper | heat |
| | light tan | copper | decomposition |
| | tan | iron | heat |
| | light golden tan | iron | decomposition |
| | yellow-tan | tin | heat |
| ***Argemone polyanthemos*** PRICKLY POPPY | | | |
| leaves | light khaki | copper | decomposition |
| roots | light yellow-tan | alum | heat |
| | tan | copper | heat |
| | medium brown | iron | heat |

| DYE SOURCE | DYE RESULT | MORDANT | PROCESSING |
|---|---|---|---|
| ***Argemone polyanthemos,*** roots, continued | | | |
| | light tan | none | heat |
| stems | light brown | iron | heat |
| ***Asclepias arenaria*** SAND MILKWEED | | | |
| leaves | light golden brown | iron | heat |
| ***Asclepias viridis*** ANTELOPE-HORN MILKWEED | | | |
| roots | light tan | tin | heat |
| ***Asimina triloba*** PAWPAW | | | |
| branches | light tan | alum | heat |
| | light golden tan | alum | decomposition |
| | medium gray-brown | iron | decomposition |
| | tan | tin | heat |
| leaves | light khaki | copper | heat |
| | light tan-peach | copper | decomposition |
| | light golden tan | iron | decomposition |
| | light yellow-tan | none | heat |
| rootstocks | light khaki | copper | heat |
| ***Baccharis salicina*** WILLOW BACCHARIS | | | |
| stems | light yellow-tan | alum | heat |
| | yellow-tan | copper | heat |
| | tan | copper | decomposition |
| | light brown | iron | decomposition |
| | light tan | none | heat |
| | light khaki | none | decomposition |
| ***Baptisia australis*** BLUE WILD-INDIGO | | | |
| leaves and upper stems | tan | alum | decomposition |
| | tan | copper | decomposition |
| | dark red-brown | iron | heat |
| | medium brown | iron | decomposition |
| | tan | tin | decomposition |
| | light khaki | none | heat |
| petals | light khaki | copper | heat |
| | light khaki | iron | decomposition |
| | light yellow-tan | none | heat |
| roots | yellow-tan | alum | heat |
| | tan | copper | heat |
| | tan | iron | heat |
| | light tan | iron | decomposition |
| | light tan | none | heat |
| stems (basal) | light tan | alum | decomposition |
| | tan | copper | heat |
| | tan | copper | decomposition |
| | dark pink-gray-brown | iron | decomposition |
| | light golden tan | tin | decomposition |
| | light tan | none | decomposition |
| ***Baptisia bracteata*** PLAINS WILD-INDIGO | | | |
| leaves | light tan | alum | decomposition |
| | tan | copper | decomposition |
| | dark brown | iron | heat |
| | medium pink-gray-brown | iron | decomposition |

| DYE SOURCE | DYE RESULT | MORDANT | PROCESSING |
|---|---|---|---|
| | light tan-peach | tin | decomposition |
| roots | light yellow-tan | alum | heat |
| | light golden tan | copper | heat |
| | yellow-tan | iron | heat |
| | light golden tan | iron | decomposition |
| | light tan | none | heat |
| ***Baptisia sphaerocarpa*** GOLDEN WILD-INDIGO | | | |
| leaves | light khaki | copper | decomposition |
| | yellow-tan | iron | heat |
| | tan | iron | decomposition |
| roots | light tan | alum | heat |
| | light golden tan | copper | heat |
| | yellow-tan | iron | heat |
| stems | light tan | alum | decomposition |
| ***Bidens aristosa*** BEARDED BEGGAR-TICKS | | | |
| flowers | dark brown | iron | heat |
| leaves | dark brown | iron | heat |
| stems | yellow-tan | alum | heat |
| | golden tan | iron | heat |
| ***Bidens bipinnata*** SPANISH NEEDLES | | | |
| heads | light yellow-tan | alum | heat |
| | yellow-tan | copper | heat |
| | tan | iron | heat |
| | light yellow-tan | none | heat |
| leaves | bright golden tan | alum | heat |
| | light tan | copper | decomposition |
| ***Brickellia eupatorioides*** FALSE BONESET | | | |
| leaves | yellow-tan | copper | decomposition |
| ***Buchnera americana*** BLUEHEARTS | | | |
| flowers | tan | copper | heat |
| | tan | iron | heat |
| leaves | brown | alum | decomposition |
| | tan | copper | decomposition |
| | olive-brown | iron | heat |
| | light tan | iron | decomposition |
| | yellow-tan | tin | decomposition |
| | light pink-brown | none | decomposition |
| ***Bumelia lanuginosa*** CHITTAMWOOD | | | |
| branchlets | light golden tan | copper | decomposition |
| leaves | tan | copper | decomposition |
| | light khaki | iron | decomposition |
| ***Callicarpa americana*** AMERICAN BEAUTYBERRY | | | |
| branchlets | light brown | alum | decomposition |
| | light brown | copper | decomposition |
| | medium brown | iron | decomposition |
| | tan | none | decomposition |
| fruits | light tan | alum | decomposition |
| | yellow-tan | copper | decomposition |
| | tan | iron | decomposition |
| leaves | tan | alum | decomposition |

| DYE SOURCE | DYE RESULT | MORDANT | PROCESSING |
|---|---|---|---|
| ***Callicarpa americana,*** leaves, continued | | | |
| | light brown | copper | decomposition |
| | light khaki | iron | heat |
| | light brown | iron | decomposition |
| | tan | none | decomposition |
| ***Calylophus hartwegii*** HARTWEG'S SUNDROP | | | |
| leaves | brown | copper | decomposition |
| | dark olive-brown | iron | heat |
| | medium gray-brown | iron | decomposition |
| petals | yellow-tan | alum | decomposition |
| | light golden tan | copper | heat |
| | tan | copper | decomposition |
| | medium gray-brown | iron | heat |
| | light brown | iron | decomposition |
| | light tan | none | decomposition |
| roots | light golden tan | alum | heat |
| | light golden tan | copper | heat |
| | medium gray-brown | iron | heat |
| | light yellow-tan | tin | heat |
| | light tan | none | heat |
| ***Calylophus serrulatus*** HALFSHRUB SUNDROP | | | |
| leaves | light yellow-tan | none | heat |
| petals | tan | copper | heat |
| roots | light tan | copper | heat |
| | medium gray-tan | iron | heat |
| stems | light tan | alum | heat |
| | light golden tan | copper | heat |
| ***Campsis radicans*** TRUMPET CREEPER | | | |
| flowers | light khaki | copper | heat |
| leaves | light tan | alum | decomposition |
| | tan | copper | decomposition |
| | light brown | iron | decomposition |
| ***Carya cordiformis*** BITTERNUT HICKORY | | | |
| branches | light tan-peach | alum | heat |
| | light tan | alum | decomposition |
| | light golden tan | copper | decomposition |
| | medium gray-brown | iron | heat |
| | tan | iron | decomposition |
| | light tan | tin | decomposition |
| | light tan | none | decomposition |
| fruits | tan | copper | decomposition |
| | medium gray-brown | iron | decomposition |
| leaves | light golden tan | copper | decomposition |
| | light brown | iron | decomposition |
| | light orange-tan | tin | decomposition |
| ***Carya illinoinensis*** PECAN | | | |
| branchlets | light orange-tan | alum | heat |
| | golden brown | copper | heat |
| | light khaki | copper | decomposition |
| | brown | iron | heat |

| DYE SOURCE | DYE RESULT | MORDANT | PROCESSING |
|---|---|---|---|
| | light golden tan | tin | heat |
| | light salmon-tan | none | heat |
| leaflets | dark brown | alum | heat |
| | light brown | alum | decomposition |
| | dark brown | copper | heat |
| | light brown | copper | decomposition |
| | dark brown | iron | heat |
| | light brown | iron | decomposition |
| | light brown | tin | heat |
| | golden brown | tin | decomposition |
| | dark brown | none | heat |
| | light brown | none | decomposition |
| leaves | medium gray-brown | iron | heat |
| ***Castanea ozarkensis*** OZARK CHINKAPIN | | | |
| branches | light tan-peach | alum | heat |
| | pink-brown | alum | decomposition |
| | tan | copper | heat |
| | tan | copper | decomposition |
| | medium gray-brown | iron | decomposition |
| | light pink-brown | none | heat |
| | light tan | none | decomposition |
| ***Castilleja indivisa*** INDIAN PAINTBRUSH | | | |
| bract tips | light tan | alum | decomposition |
| | light yellow-tan | iron | heat |
| | light tan | iron | decomposition |
| | light tan | none | decomposition |
| leaves and stems | yellow-tan | alum | decomposition |
| | yellow-tan | copper | decomposition |
| | yellow-tan | none | decomposition |
| ***Castilleja purpurea*** **var. *citrina*** YELLOW PAINTBRUSH | | | |
| leaves | light tan | alum | decomposition |
| | light golden tan | copper | decomposition |
| | tan | iron | decomposition |
| petals and bracts | light yellow-tan | alum | decomposition |
| | light tan | copper | decomposition |
| | light golden brown | iron | heat |
| | yellow-tan | iron | decomposition |
| ***Catalpa speciosa*** NORTHERN CATALPA | | | |
| branches | cinnamon-brown | alum | heat |
| | dark cinnamon-brown | copper | heat |
| | light golden tan | copper | decomposition |
| | cinnamon-brown | iron | heat |
| | cinnamon-brown | none | heat |
| leaves | cinnamon-brown | alum | heat |
| | yellow-tan | alum | decomposition |
| | brown | copper | heat |
| | golden tan | copper | decomposition |
| | light golden brown | iron | decomposition |
| | brown | none | heat |

| DYE SOURCE | DYE RESULT | MORDANT | PROCESSING |
|---|---|---|---|
| ***Ceanothus americanus*** NEW JERSEY TEA | | | |
| leaves | light khaki | copper | heat |
| | light khaki | iron | decomposition |
| rhizomes and roots | medium pink-tan | alum | heat |
| | yellow-tan | copper | heat |
| | light tan | copper | decomposition |
| | light pink-brown | none | heat |
| stems | light tan | copper | heat |
| | light golden tan | copper | decomposition |
| | medium gray-brown | iron | heat |
| | light tan-peach | none | heat |
| ***Cephalanthus occidentalis*** BUTTONBUSH | | | |
| branches | light tan-peach | alum | heat |
| | light pink-brown | alum | decomposition |
| | light tan | copper | heat |
| | tan | iron | decomposition |
| | yellow-tan | tin | heat |
| | medium pink-brown | none | decomposition |
| flowers | light khaki | iron | heat |
| ***Cercis canadensis*** REDBUD | | | |
| bark of branches | yellow-tan | copper | heat |
| | light tan-peach | none | heat |
| bark of trunk | light tan | alum | decomposition |
| | light tan | copper | decomposition |
| | yellow-tan | iron | decomposition |
| roots | light yellow-tan | alum | heat |
| | light tan | copper | heat |
| | medium gray-brown | copper | decomposition |
| | medium brown | iron | decomposition |
| | light yellow-tan | tin | heat |
| | pink-brown | tin | decomposition |
| | light tan-peach | none | heat |
| | light tan-peach | none | decomposition |
| ***Cicuta macula*** WATER-HEMLOCK | | | |
| flowers | medium brown | iron | heat |
| | light khaki | iron | decomposition |
| | yellow-tan | none | heat |
| leaves | olive-brown | iron | heat |
| ***Cirsium undulatum*** WAVY-LEAF THISTLE | | | |
| pappus | light khaki | copper | heat |
| roots | light tan | alum | heat |
| | tan | copper | heat |
| | golden tan | iron | heat |
| | golden tan | tin | heat |
| ***Comandra umbellata*** BASTARD TOAD-FLAX | | | |
| leaves | light yellow-tan | alum | heat |
| | golden tan | copper | heat |
| | dark olive-brown | iron | heat |
| | light brown | iron | decomposition |
| | light tan-peach | none | decomposition |

| DYE SOURCE | DYE RESULT | MORDANT | PROCESSING |
|---|---|---|---|
| stems | golden tan | copper | heat |
| | olive-brown | iron | heat |
| ***Coreopsis tinctoria*** PLAINS TICKSEED | | | |
| disk florets and bracts | light golden tan | alum | decomposition |
| | golden brown | copper | heat |
| | golden tan | copper | decomposition |
| | black-brown | iron | heat |
| | olive-brown | iron | decomposition |
| | yellow-tan | none | decomposition |
| ray florets | golden tan | alum | decomposition |
| | dark brown | iron | heat |
| | light golden brown | iron | decomposition |
| | bright golden tan | tin | decomposition |
| ***Cornus drummondii*** ROUGH-LEAF DOGWOOD | | | |
| branches | light tan | none | heat |
| branchlets | yellow-tan | alum | heat |
| | light tan-peach | alum | decomposition |
| | tan | copper | decomposition |
| | light khaki | iron | decomposition |
| fruits | tan | copper | decomposition |
| leaves | golden tan | copper | decomposition |
| | dark gray-brown | iron | heat |
| | olive-brown | iron | decomposition |
| | light golden tan | tin | decomposition |
| ***Cornus florida*** FLOWERING DOGWOOD | | | |
| bark | yellow-tan | alum | heat |
| | golden tan | copper | heat |
| | light yellow-tan | copper | decomposition |
| | yellow-tan | iron | decomposition |
| | light yellow-tan | none | heat |
| fruits | light khaki | copper | decomposition |
| leaves | yellow-tan | alum | heat |
| | light golden brown | copper | heat |
| | golden tan | copper | decomposition |
| | black-brown | iron | heat |
| | light tan-peach | none | heat |
| | light golden tan | none | decomposition |
| ***Cotinus obovatus*** SMOKE-TREE | | | |
| bark | dark olive-brown | iron | heat |
| leaves | light yellow-tan | alum | decomposition |
| | light khaki | copper | heat |
| | golden tan | copper | decomposition |
| | black-brown | iron | heat |
| wood | orange-brown | copper | heat |
| | dark brown | iron | heat |
| ***Cucurbita foetidissima*** BUFFALO GOURD | | | |
| leaves | golden tan | copper | heat |
| | dark gray-brown | iron | heat |
| | light brown | iron | decomposition |
| | yellow-tan | tin | heat |

| DYE SOURCE | DYE RESULT | MORDANT | PROCESSING |
|---|---|---|---|
| ***Cucurbita foetidissima,*** leaves, continued | | | |
| | light yellow-tan | tin | decomposition |
| stems | light tan | copper | decomposition |
| | light tan | none | decomposition |
| ***Cyperus squarrosus*** BEARDED FLATSEDGE | | | |
| leaves | light tan | alum | heat |
| | light golden tan | copper | heat |
| | golden tan | iron | heat |
| spikes | light golden tan | copper | heat |
| | light golden tan | copper | decomposition |
| ***Dalea candida*** WHITE PRAIRIE-CLOVER | | | |
| roots | light yellow-tan | alum | heat |
| | yellow-tan | copper | heat |
| | medium brown | iron | heat |
| | orange-tan | tin | heat |
| | light yellow-tan | none | heat |
| ***Dalea lanata*** WOOLLY DALEA | | | |
| flowers | yellow-tan | tin | decomposition |
| roots | light yellow-tan | copper | heat |
| | light tan | iron | heat |
| | light tan | iron | decomposition |
| ***Dalea villosa*** SILKY PRAIRIE-CLOVER | | | |
| flowers | light khaki | iron | decomposition |
| leaves | dark olive-brown | iron | heat |
| stems | tan | iron | heat |
| ***Desmanthus illinoensis*** ILLINOIS BUNDLEFLOWER | | | |
| fruits | dark gray-brown | iron | heat |
| stems | light golden tan | copper | decomposition |
| ***Desmodium glutinosum*** STICKY TICKCLOVER | | | |
| leaves | light khaki | iron | heat |
| stems | light tan | copper | decomposition |
| | light golden tan | iron | decomposition |
| | light tan | tin | heat |
| ***Diospyros virginiana*** PERSIMMON | | | |
| bark | light tan | alum | heat |
| | light golden tan | copper | heat |
| | light yellow-tan | tin | heat |
| | light tan | none | heat |
| branchlets | light golden tan | copper | decomposition |
| | golden tan | iron | heat |
| | light yellow-tan | tin | decomposition |
| | yellow-tan | none | heat |
| leaves | golden tan | copper | heat |
| ***Dracopis amplexicaulis*** CLASPING-LEAVED CONEFLOWER | | | |
| disk florets | dark brown | tin | heat |
| leaves | light golden tan | alum | decomposition |
| | light tan-peach | copper | decomposition |
| | light tan | iron | decomposition |
| stems | light golden tan | copper | decomposition |

| DYE SOURCE | DYE RESULT | MORDANT | PROCESSING |
|---|---|---|---|
| ***Echinacea angustifolia*** PURPLE PRAIRIE CONEFLOWER | | | |
| disk florets | light golden tan | alum | decomposition |
| | light khaki | copper | heat |
| | light brown | copper | decomposition |
| | tan | iron | decomposition |
| leaves | light khaki | copper | heat |
| | light khaki | copper | decomposition |
| ***Elephantopus carolinianus*** LEAFY ELEPHANT'S FOOT | | | |
| leaves | light khaki | copper | decomposition |
| stems | light khaki | alum | decomposition |
| ***Erigeron strigosus*** DAISY FLEABANE | | | |
| disk florets | dark olive-brown | iron | heat |
| ***Euphorbia marginata*** SNOW-ON-THE-MOUNTAIN | | | |
| stems | light tan | none | decomposition |
| ***Eustoma exaltatum*** PRAIRIE GENTIAN | | | |
| leaves | light yellow-tan | alum | heat |
| | light yellow-tan | alum | decomposition |
| | light tan | copper | heat |
| | light yellow-tan | copper | decomposition |
| | medium pink-brown | iron | heat |
| | light yellow-tan | iron | decomposition |
| | light tan-peach | none | heat |
| | yellow-tan | none | decomposition |
| petals | light tan | alum | heat |
| | golden tan | copper | heat |
| | golden tan | iron | heat |
| | light tan | none | heat |
| roots | light yellow-tan | alum | heat |
| | tan | copper | heat |
| | light tan | iron | heat |
| | light tan | none | heat |
| stems | light tan-peach | alum | heat |
| | light tan | copper | heat |
| | light tan | iron | heat |
| | light tan | iron | decomposition |
| | light tan | none | heat |
| ***Fraxinus americana*** WHITE ASH | | | |
| fruits | light golden tan | alum | decomposition |
| | light golden tan | copper | decomposition |
| | tan | iron | decomposition |
| | tan | none | decomposition |
| ***Fraxinus pennsylvanica*** GREEN ASH | | | |
| leaves | dark gray-brown | iron | heat |
| ***Fraxinus quadrangulata*** BLUE ASH | | | |
| bark | light khaki | copper | decomposition |
| ***Froelichia floridana*** FIELD SNAKE-COTTON | | | |
| flowers | golden tan | copper | heat |
| | medium brown | iron | heat |
| | light tan | none | heat |

| DYE SOURCE | DYE RESULT | MORDANT | PROCESSING |
|---|---|---|---|
| ***Froelichia floridana,*** continued | | | |
| leaves | light khaki | copper | heat |
| | medium brown | iron | heat |
| stems | tan | copper | heat |
| | medium brown | iron | heat |
| ***Gaillardia aestivalis*** PRAIRIE GAILLARDIA | | | |
| leaves | dark olive-brown | iron | heat |
| ray florets | light khaki | copper | heat |
| ***Glandularia canadensis*** ROSE VERVAIN | | | |
| flowers | light yellow-tan | alum | decomposition |
| | light yellow-tan | copper | decomposition |
| | light golden tan | iron | heat |
| | light yellow-tan | iron | decomposition |
| | light yellow-tan | none | decomposition |
| leaves and stems | light yellow-tan | alum | decomposition |
| | yellow-tan | copper | decomposition |
| | yellow-tan | iron | decomposition |
| | light golden brown | tin | decomposition |
| | light yellow-tan | none | decomposition |
| ***Gymnocladus dioicus*** KENTUCKY COFFEE-TREE | | | |
| bark | light khaki | iron | heat |
| | light yellow-tan | none | heat |
| ***Hedyotis nigricans*** PRAIRIE BLUETS | | | |
| leaves and stems | tan | alum | heat |
| | light khaki | alum | decomposition |
| | light brown | copper | heat |
| | golden tan | copper | decomposition |
| | medium brown | iron | heat |
| | golden tan | iron | decomposition |
| | light tan | none | heat |
| | tan | none | decomposition |
| roots | light tan | copper | heat |
| | light tan | iron | heat |
| | light tan-peach | tin | heat |
| ***Helenium amarum*** BITTER SNEEZEWEED | | | |
| heads | light khaki | copper | decomposition |
| stems | yellow-tan | copper | decomposition |
| ***Helianthus annuus*** ANNUAL SUNFLOWER | | | |
| leaves | light brown | copper | decomposition |
| | dark brown | iron | heat |
| | light khaki | iron | decomposition |
| ***Helianthus maximiliani*** MAXIMILIAN'S SUNFLOWER | | | |
| disk florets and phyllaries | light golden tan | copper | decomposition |
| | tan | iron | decomposition |
| | light tan | none | decomposition |
| leaves | light khaki | copper | decomposition |
| | medium brown | iron | heat |
| | light khaki | iron | decomposition |
| rhizomes | light brown | copper | decomposition |
| | medium brown | iron | heat |

| DYE SOURCE | DYE RESULT | MORDANT | PROCESSING |
|---|---|---|---|
| | medium brown | iron | decomposition |
| | medium brown | none | decomposition |
| stems | light brown | copper | decomposition |
| ***Helianthus tuberosus*** JERUSALEM ARTICHOKE | | | |
| leaves | light brown | copper | decomposition |
| stems | light khaki | alum | decomposition |
| | tan | copper | decomposition |
| | light brown | iron | decomposition |
| ***Hibiscus laevis*** SCARLET ROSE MALLOW | | | |
| leaves | tan | copper | heat |
| | light golden tan | copper | decomposition |
| | light khaki | iron | heat |
| | light golden tan | iron | decomposition |
| petals | olive-brown | iron | heat |
| | light khaki | iron | decomposition |
| ***Ipomoea leptophylla*** BUSH MORNING-GLORY | | | |
| leaves | light tan | alum | decomposition |
| | light tan | copper | decomposition |
| | dark gray-brown | iron | decomposition |
| | light golden tan | tin | decomposition |
| | light tan | none | decomposition |
| stems | light khaki | iron | heat |
| ***Ipomopsis rubra*** STANDING CYPRESS | | | |
| leaves | light khaki | iron | decomposition |
| petals | light golden tan | iron | heat |
| stems | light khaki | iron | heat |
| ***Juglans nigra*** BLACK WALNUT | | | |
| bark | yellow-tan | alum | heat |
| | yellow-tan | copper | heat |
| | medium gray-brown | iron | heat |
| | yellow-tan | tin | heat |
| | light salmon-tan | none | heat |
| fruits | light tan-peach | alum | heat |
| | light golden tan | copper | heat |
| | light tan-peach | iron | heat |
| | light tan | tin | heat |
| involucral husks | dark brown | alum | heat |
| | pink-brown | alum | decomposition |
| | dark brown | copper | heat |
| | cinnamon-brown | copper | decomposition |
| | black-brown | iron | heat |
| | dark red-brown | iron | decomposition |
| | cinnamon-brown | tin | heat |
| | dark brown | none | heat |
| | orange-brown | none | decomposition |
| leaves | golden tan | alum | heat |
| | light golden tan | alum | decomposition |
| | light brown | copper | heat |
| | yellow-tan | copper | decomposition |
| | dark gray-brown | iron | heat |

| DYE SOURCE | DYE RESULT | MORDANT | PROCESSING |
|---|---|---|---|
| ***Juglans nigra,*** leaves, continued | | | |
| | golden brown | iron | decomposition |
| | orange-tan | tin | decomposition |
| | light cinnamon-brown | none | heat |
| | yellow-tan | none | decomposition |
| ***Juniperus virginiana*** EASTERN RED CEDAR | | | |
| bark | light pink-brown | copper | heat |
| | light tan | iron | decomposition |
| | light orange-tan | tin | heat |
| roots | light tan-peach | alum | heat |
| | light tan | copper | heat |
| | light tan-peach | none | heat |
| ***Liatris punctata*** DOTTED GAYFEATHER | | | |
| leaves and stems | light khaki | copper | decomposition |
| | olive-brown | iron | heat |
| phyllaries | light khaki | copper | decomposition |
| | light khaki | iron | heat |
| | light khaki | iron | decomposition |
| ***Linum sulcatum*** GROOVED FLAX | | | |
| leaves, stems, and petals | medium brown | iron | heat |
| ***Liquidambar styraciflua*** SWEET GUM | | | |
| bark of branches | yellow-tan | alum | heat |
| | light golden brown | copper | heat |
| | medium gray-brown | iron | decomposition |
| | light golden tan | tin | heat |
| | orange-tan | tin | decomposition |
| | light tan | none | heat |
| | light yellow-tan | none | decomposition |
| bark of trunk | tan | copper | decomposition |
| leaves | golden tan | alum | heat |
| | golden tan | copper | decomposition |
| | dark gray-brown | iron | heat |
| | dark brown | iron | decomposition |
| | tan | none | heat |
| | yellow-tan | none | decomposition |
| roots | brown | copper | decomposition |
| | dark gray-brown | iron | decomposition |
| | light tan | none | decomposition |
| ***Maclura pomifera*** OSAGE ORANGE | | | |
| bark | golden tan | copper | heat |
| | yellow-tan | none | heat |
| leaves | olive-brown | iron | heat |
| | light khaki | iron | decomposition |
| roots | bright golden tan | alum | heat |
| | dark brown | iron | heat |
| ***Mentzelia nuda*** BRACTLESS BLAZING-STAR | | | |
| leaves | light khaki | copper | heat |
| roots | tan | alum | heat |
| | light tan | copper | heat |
| | light cinnamon-brown | iron | heat |

| DYE SOURCE | DYE RESULT | MORDANT | PROCESSING |
|---|---|---|---|
| | light tan | none | heat |
| stems | tan | copper | heat |
| | light tan | iron | heat |
| | light golden tan | tin | heat |
| ***Monarda fistulosa*** WILD BERGAMOT | | | |
| petals | light golden tan | alum | decomposition |
| | yellow-tan | copper | decomposition |
| | light golden brown | iron | heat |
| | light golden tan | iron | decomposition |
| ***Morus rubra*** RED MULBERRY | | | |
| bark | light tan | alum | decomposition |
| | light golden tan | none | decomposition |
| ***Neptunia lutea*** YELLOW NEPTUNE | | | |
| leaves | olive-brown | iron | heat |
| ***Nuttallanthus canadensis*** TOAD-FLAX | | | |
| leaves and stems | tan | alum | heat |
| | medium brown | copper | heat |
| | medium brown | iron | heat |
| | brown | none | heat |
| petals | golden tan | alum | heat |
| | tan | copper | heat |
| | yellow-tan | iron | heat |
| | light yellow-tan | none | heat |
| ***Nyssa sylvatica*** BLACK GUM | | | |
| bark | light khaki | copper | heat |
| | tan | iron | decomposition |
| | light tan | none | heat |
| | light golden tan | none | decomposition |
| branchlets | light golden tan | copper | decomposition |
| | light golden tan | none | decomposition |
| leaves | light yellow-tan | alum | decomposition |
| | yellow-tan | copper | decomposition |
| ***Oenothera heterophylla*** SAND EVENING-PRIMROSE | | | |
| flowers | light khaki | copper | heat |
| | golden tan | copper | decomposition |
| | dark gray-brown | iron | decomposition |
| | light yellow-tan | none | heat |
| leaves | light golden brown | copper | decomposition |
| | dark brown | iron | heat |
| | dark brown | iron | decomposition |
| | light yellow-tan | none | heat |
| roots | light khaki | copper | heat |
| | tan | copper | decomposition |
| | dark brown | iron | decomposition |
| | light tan | none | heat |
| | light tan | none | decomposition |
| stems | light yellow-tan | alum | decomposition |
| | light khaki | copper | heat |
| | tan | copper | decomposition |
| | light yellow-tan | none | heat |

| DYE SOURCE | DYE RESULT | MORDANT | PROCESSING |
|---|---|---|---|
| ***Oenothera macrocarpa*** BIGFRUIT EVENING-PRIMROSE | | | |
| fruits immature | light yellow-tan | alum | decomposition |
| | light brown | copper | heat |
| | light golden tan | copper | decomposition |
| | light tan | none | decomposition |
| leaves | golden tan | alum | decomposition |
| | light khaki | copper | heat |
| | light khaki | copper | decomposition |
| | dark gray-brown | iron | heat |
| | light tan | none | heat |
| | light tan | none | decomposition |
| petals | light khaki | copper | decomposition |
| roots | tan | copper | heat |
| stems | light golden tan | alum | decomposition |
| | light golden tan | copper | heat |
| | light golden tan | copper | decomposition |
| | light brown | iron | decomposition |
| | light golden tan | tin | decomposition |
| | light tan | none | decomposition |
| ***Packera obovata*** ROUNDLEAF GROUNDSEL | | | |
| leaves and stems | light golden tan | iron | decomposition |
| ***Parthenocissus quinquefolia*** VIRGINIA CREEPER | | | |
| bark | golden tan | alum | heat |
| | medium brown | copper | heat |
| | tan | copper | decomposition |
| | light tan | none | heat |
| leaves | light khaki | iron | heat |
| ***Pediomelum cuspidatum*** TALL-BREAD SCURFPEA | | | |
| leaves | light golden tan | alum | decomposition |
| | tan | copper | decomposition |
| | light tan | tin | decomposition |
| roots | light tan | tin | heat |
| ***Phlox pilosa*** PRAIRIE PHLOX | | | |
| stems | light tan | iron | decomposition |
| ***Phyla lanceolata*** LANCELEAF FROG-FRUIT | | | |
| leaves | light khaki | iron | heat |
| stems | light tan | iron | decomposition |
| ***Platanus occidentalis*** SYCAMORE | | | |
| bark of branches | yellow-tan | alum | heat |
| | yellow-tan | copper | heat |
| | light tan | copper | decomposition |
| | light brown | iron | heat |
| | yellow-tan | iron | decomposition |
| | light yellow-tan | tin | decomposition |
| | light tan | none | heat |
| ***Podophyllum peltatum*** MAY-APPLE | | | |
| leaves | light tan | alum | decomposition |
| | light tan | copper | decomposition |
| | light tan-peach | none | decomposition |
| rhizomes and roots | medium brown | iron | heat |
| | light golden tan | iron | decomposition |

| DYE SOURCE | DYE RESULT | MORDANT | PROCESSING |
|---|---|---|---|
| ***Polygala alba*** WHITE MILKWORT | | | |
| leaves and stems | golden tan | iron | heat |
| ***Polygonum pensylvanicum*** PENNSYLVANIA SMARTWEED | | | |
| flowers | light golden brown | iron | decomposition |
| stems | light khaki | iron | decomposition |
| ***Polygonum punctatum*** DOTTED SMARTWEED | | | |
| flowers | light khaki | copper | decomposition |
| | light yellow-tan | tin | decomposition |
| ***Polytaenia nuttallii*** PRAIRIE PARSLEY | | | |
| flowers | light brown | copper | decomposition |
| | yellow-tan | iron | decomposition |
| roots | light khaki | copper | decomposition |
| ***Populus deltoides*** EASTERN COTTONWOOD | | | |
| roots | tan | copper | heat |
| | tan | copper | decomposition |
| | light yellow-tan | tin | heat |
| stems | olive-brown | iron | heat |
| ***Potentilla arguta*** TALL CINQUEFOIL | | | |
| flowers | dark brown | iron | heat |
| | medium gray-brown | iron | decomposition |
| | golden tan | none | heat |
| leaves | light yellow-tan | alum | decomposition |
| | yellow-tan | copper | decomposition |
| | tan | iron | decomposition |
| | light yellow-tan | tin | decomposition |
| | light salmon-tan | none | decomposition |
| rootstocks | light tan | copper | heat |
| | dark gray-brown | iron | heat |
| | light tan-peach | none | heat |
| stems | black-brown | iron | heat |
| | light yellow-tan | none | heat |
| ***Prosopis glandulosa*** MESQUITE | | | |
| roots | light yellow-tan | alum | heat |
| | tan | copper | heat |
| | yellow-tan | iron | heat |
| | light tan | iron | decomposition |
| | light yellow-tan | none | heat |
| ***Psoralidium tenuiflorum*** SLIM-FLOWER SCURFPEA | | | |
| leaves | dark olive-brown | iron | heat |
| | olive-brown | iron | decomposition |
| roots | light tan | alum | heat |
| | light golden tan | copper | heat |
| | light golden tan | iron | heat |
| | yellow-tan | tin | heat |
| ***Ptilimnium nuttallii*** MOCK BISHOP'S-WEED | | | |
| leaves and stems | medium brown | iron | heat |
| roots | light khaki | iron | heat |
| ***Pyrrhopappus grandiflorus*** MORNING STAR | | | |
| ligulate florets | light brown | iron | heat |

| DYE SOURCE | DYE RESULT | MORDANT | PROCESSING |
|---|---|---|---|
| ***Ratibida columnifera*** MEXICAN HAT | | | |
| leaves and stems | tan | copper | decomposition |
| | medium brown | iron | heat |
| | tan | iron | decomposition |
| ***Rhus copallinum*** WINGED SUMAC | | | |
| bark | yellow-tan | alum | heat |
| | light tan | alum | decomposition |
| | light golden tan | copper | decomposition |
| | black-brown | iron | heat |
| | golden tan | tin | heat |
| | light yellow-tan | none | heat |
| flowers | light orange-tan | alum | decomposition |
| | dark gray-brown | iron | heat |
| | light orange-tan | tin | decomposition |
| leaves | light brown | copper | heat |
| | dark gray-brown | iron | heat |
| ***Rhus glabra*** SMOOTH SUMAC | | | |
| bark | golden tan | alum | heat |
| | olive-brown | copper | heat |
| | black-brown | iron | heat |
| | light khaki | none | heat |
| fruits | golden brown | alum | heat |
| | light pink-brown | copper | decomposition |
| | medium gray-brown | iron | heat |
| | dark brown | tin | heat |
| | dark gray-brown | tin | decomposition |
| | yellow-tan | none | heat |
| | medium pink-tan | none | decomposition |
| leaves | dark gray-brown | iron | heat |
| ***Robinia hispida*** BRISTLY LOCUST | | | |
| flowers | light khaki | iron | heat |
| | light tan | none | heat |
| leaves | olive-brown | iron | heat |
| stems | light golden tan | alum | heat |
| | tan | copper | heat |
| | light brown | iron | heat |
| | light yellow-tan | none | heat |
| ***Robinia pseudoacacia*** BLACK LOCUST | | | |
| bark | light tan | copper | heat |
| | light tan | iron | heat |
| | light yellow-tan | tin | heat |
| petals | yellow-tan | iron | heat |
| ***Rudbeckia grandiflora*** ROUGH CONEFLOWER | | | |
| disk florets | light golden tan | alum | decomposition |
| | golden tan | copper | heat |
| | light tan | copper | decomposition |
| | light tan-peach | none | decomposition |
| leaves and stems | light golden tan | alum | decomposition |
| | tan | copper | decomposition |
| | light brown | iron | heat |
| | tan | iron | decomposition |

| DYE SOURCE | DYE RESULT | MORDANT | PROCESSING |
| --- | --- | --- | --- |
| | light golden tan | none | decomposition |
| ray florets | light tan | copper | decomposition |
| ***Rudbeckia hirta*** BLACK-EYED SUSAN | | | |
| disk florets | medium brown | tin | heat |
| | golden brown | tin | decomposition |
| leaves | light brown | copper | decomposition |
| ray florets | light tan | copper | decomposition |
| roots | light tan | copper | decomposition |
| | tan | tin | heat |
| ***Rumex altissimus*** PALE DOCK | | | |
| flowers | light khaki | copper | decomposition |
| | dark olive-brown | iron | heat |
| leaves | light khaki | iron | decomposition |
| stems | olive-brown | iron | heat |
| ***Sabatia campestris*** PRAIRIE ROSE GENTIAN | | | |
| leaves and stems | light yellow-tan | alum | heat |
| | light golden tan | copper | heat |
| | yellow-tan | iron | heat |
| petals | light yellow-tan | copper | decomposition |
| ***Salix caroliniana*** CAROLINA WILLOW | | | |
| branches | light pink-brown | copper | heat |
| | light tan-peach | copper | decomposition |
| | medium gray-brown | iron | heat |
| | light tan | iron | decomposition |
| ***Salix exigua*** SANDBAR WILLOW | | | |
| stems | light golden tan | copper | heat |
| | light tan | copper | decomposition |
| ***Salix nigra*** BLACK WILLOW | | | |
| bark | light golden tan | copper | heat |
| branchlets | light golden brown | copper | heat |
| | olive-brown | iron | heat |
| ***Sambucus canadensis*** ELDERBERRY | | | |
| bark | light tan | alum | decomposition |
| | tan | copper | heat |
| | tan | copper | decomposition |
| | light tan | none | decomposition |
| leaves | dark olive-brown | iron | heat |
| ***Sanguinaria canadensis*** BLOODROOT | | | |
| leaves and stems | light golden tan | copper | decomposition |
| | yellow-tan | iron | decomposition |
| rhizomes and roots | light orange-tan | alum | heat |
| | light orange-tan | copper | heat |
| | light yellow-tan | copper | decomposition |
| | orange-tan | iron | heat |
| | light yellow-tan | iron | decomposition |
| ***Sapindus drummondii*** SOAPBERRY | | | |
| flowers | medium brown | iron | decomposition |
| leaves | light brown | iron | heat |
| | yellow-tan | iron | decomposition |
| stems | tan | copper | decomposition |

| DYE SOURCE | DYE RESULT | MORDANT | PROCESSING |
|---|---|---|---|
| ***Sassafras albidum*** SASSAFRAS | | | |
| bark | light yellow-tan | alum | heat |
| | yellow-tan | copper | heat |
| | medium brown | iron | heat |
| | light tan-peach | none | heat |
| | light yellow-tan | none | decomposition |
| flowers | olive-brown | iron | heat |
| roots | light tan | copper | heat |
| | light yellow-tan | tin | heat |
| ***Sedum nuttallianum*** YELLOW STONECROP | | | |
| fruits, flowers, and stems | yellow-tan | iron | heat |
| ***Senecio riddellii*** SAND GROUNDSEL | | | |
| disk and ray florets | light golden tan | alum | heat |
| | light golden tan | copper | heat |
| | light brown | iron | heat |
| | light tan-peach | iron | decomposition |
| | light yellow-tan | none | heat |
| leaves | light tan | alum | decomposition |
| | light khaki | copper | heat |
| | light tan | copper | decomposition |
| | light golden brown | iron | heat |
| | light golden tan | iron | decomposition |
| roots | light golden tan | alum | heat |
| | light golden tan | copper | heat |
| | tan | iron | heat |
| | yellow-tan | tin | heat |
| | light tan | tin | decomposition |
| | light tan | none | heat |
| ***Silphium integrifolium*** PRAIRIE ROSINWEED | | | |
| leaves | tan | copper | decomposition |
| | dark brown | iron | heat |
| roots | light khaki | alum | decomposition |
| | light khaki | copper | heat |
| | light khaki | copper | decomposition |
| | light brown | none | heat |
| ***Sisyrinchium angustifolium*** BLUE-EYED GRASS | | | |
| flowers | golden tan | iron | decomposition |
| | light golden brown | tin | decomposition |
| ***Smilax bona-nox*** GREENBRIER | | | |
| leaves and stems | light golden tan | copper | heat |
| | light khaki | iron | heat |
| ***Solanum dimidiatum*** WESTERN HORSE-NETTLE | | | |
| leaves and stems | bright golden tan | tin | heat |
| | black-brown | iron | heat |
| roots | light khaki | copper | heat |
| | light khaki | iron | decomposition |
| ***Solanum elaeagnifolium*** SILVERLEAF NIGHTSHADE | | | |
| leaves | light khaki | copper | heat |
| | olive-brown | iron | heat |

| DYE SOURCE | DYE RESULT | MORDANT | PROCESSING |
|---|---|---|---|
| ***Solidago missouriensis*** PLAINS GOLDENROD | | | |
| heads | bright golden tan | tin | heat |
| | black-brown | iron | heat |
| leaves | black-brown | iron | heat |
| rhizomes and roots | light khaki | copper | heat |
| ***Solidago rigida*** STIFF GOLDENROD | | | |
| stems | light golden tan | alum | decomposition |
| | golden tan | copper | decomposition |
| | tan | iron | decomposition |
| | light golden tan | none | decomposition |
| ***Sorghastrum nutans*** INDIANGRASS | | | |
| leaves | light golden tan | copper | decomposition |
| spikelets | light khaki | copper | heat |
| ***Stillingia sylvatica*** QUEEN'S DELIGHT | | | |
| leaves | dark brown | iron | heat |
| | golden tan | none | heat |
| pistillate flowers | dark brown | iron | heat |
| | golden tan | none | heat |
| roots | light golden tan | alum | heat |
| | light tan | alum | decomposition |
| | light golden brown | copper | heat |
| | medium brown | iron | heat |
| | medium gray-brown | iron | decomposition |
| | light yellow-tan | tin | decomposition |
| | light tan | none | heat |
| | light tan-peach | none | heat |
| staminate flowers | dark brown | iron | heat |
| | medium brown | iron | decomposition |
| | golden tan | none | heat |
| stems | dark gray-brown | iron | heat |
| ***Teucrium canadense*** AMERICAN GERMANDER | | | |
| leaves | light tan | alum | decomposition |
| | tan | copper | decomposition |
| | light tan | tin | decomposition |
| ***Thelesperma filifolium*** GREENTHREAD | | | |
| leaves | bright golden tan | alum | decomposition |
| | cinnamon-brown | copper | decomposition |
| | dark olive-brown | iron | heat |
| | dark olive-brown | iron | decomposition |
| ray florets | orange-tan | iron | heat |
| | bright golden tan | iron | decomposition |
| | bright orange-tan | tin | decomposition |
| ***Ulmus rubra*** SLIPPERY ELM | | | |
| bark | light tan | alum | decomposition |
| | tan | copper | heat |
| | light tan | none | decomposition |
| leaves | light yellow-tan | alum | heat |
| | light tan | alum | decomposition |
| | light orange-tan | copper | heat |

| DYE SOURCE | DYE RESULT | MORDANT | PROCESSING |
|---|---|---|---|
| ***Ulmus rubra,*** leaves, continued | | | |
| | tan | copper | decomposition |
| | dark olive-brown | iron | heat |
| | light khaki | iron | decomposition |
| | light tan-peach | none | heat |
| ***Vaccinium arboreum*** FARKLEBERRY | | | |
| leaves | yellow-tan | alum | decomposition |
| | golden tan | copper | decomposition |
| | dark gray-brown | iron | heat |
| | light brown | iron | decomposition |
| | bright golden tan | tin | heat |
| | light tan | none | decomposition |
| rootstocks | light tan-peach | iron | heat |
| stems | light golden tan | copper | heat |
| | light yellow-tan | tin | heat |
| | light yellow-tan | none | heat |
| ***Vernonia baldwinii*** WESTERN IRONWEED | | | |
| stems | tan | alum | decomposition |
| | light brown | copper | decomposition |
| | light brown | iron | decomposition |
| | tan | none | decomposition |
| ***Viburnum rufidulum*** RUSTY BLACK HAW | | | |
| bark | light salmon-tan | alum | heat |
| | light yellow-tan | alum | decomposition |
| | light pink-brown | copper | heat |
| | light yellow-tan | copper | decomposition |
| | medium brown | iron | heat |
| | light orange-tan | iron | decomposition |
| | golden tan | tin | decomposition |
| | light yellow-tan | none | decomposition |
| flowers | light golden tan | copper | decomposition |
| | light tan | none | decomposition |
| leaves | light khaki | copper | heat |
| | light golden tan | copper | decomposition |
| | light khaki | iron | decomposition |
| pedicels | light tan | copper | heat |
| | light golden tan | copper | decomposition |
| | light brown | iron | decomposition |
| roots | light tan | iron | heat |
| | light golden tan | iron | decomposition |
| | light golden tan | tin | heat |
| ***Vitis aestivalis*** PIGEON GRAPE | | | |
| bark | light tan-peach | alum | heat |
| | tan | copper | heat |
| | light tan | none | heat |

# Nature's Neutral Dye Color

Our search for natural dyes resulted in various black colors, shown here with Munsell notations. Following these color chips is a list of the plants that were used to obtain these colors, including mordant and processing conditions.

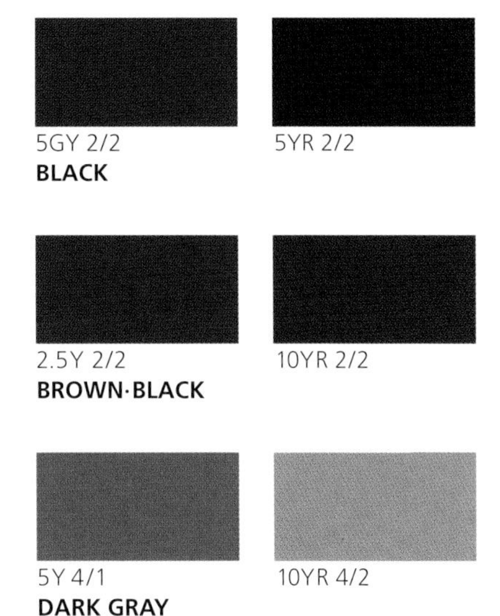

7.5Y 3/2

7.5Y 4/2

**DARK OLIVE·GRAY**

5YR 4/1

**DARK ROSE·GRAY**

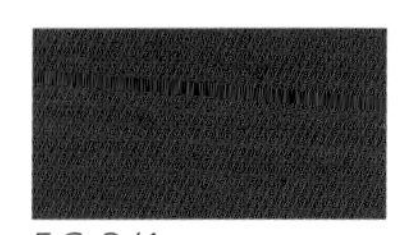

5G 3/1

**DARK TEAL·GRAY**

10Y 6/4

**GREEN·GRAY**

5Y 8/2

**LIGHT BEIGE·GRAY**

10GY 8/1

**LIGHT BLUE·GRAY**

5GY 8/1

2.5Y 7/2

2.5Y 8/2

5Y 7/1

**LIGHT GRAY**

5Y 9/1 10YR 7/2 10YR 8/1

**LIGHT GRAY, continued**

10Y 7/2

**LIGHT GREEN·GRAY**

5Y 7/2

**LIGHT KHAKI·GRAY**

7.5Y 7/2

**LIGHT OLIVE·GRAY**

5YR 8/2 7.5YR 8/2

**LIGHT PINK·GRAY**

10GY 7/1

**MEDIUM BLUE·GRAY**

5GY 6/1 10Y 6/2

**MEDIUM GREEN·GRAY**

2.5Y 6/2

5Y 5/1

5Y 6/1

10YR 5/1

**MEDIUM GRAY**

10YR 6/1

**MEDIUM GRAY, continued**

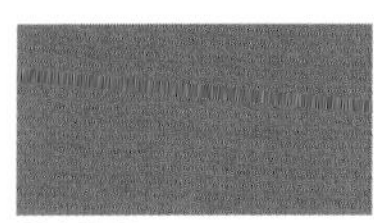
5Y 5/2

5Y 6/2

**MEDIUM KHAKI·GRAY**

7.5Y 5/2

7.5Y 6/2

**MEDIUM OLIVE·GRAY**

5YR 6/1

**MEDIUM PINK·GRAY**

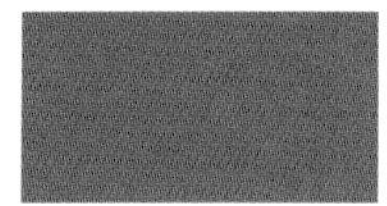
10Y 2/2

**OLIVE·BLACK**

5Y 4/2

**OLIVE·GRAY**

2.5GY 8/2
**PALE GREEN·GRAY**

10R 8/2
**PINK·GRAY**

2.5YR 9/2

5RP 2/1
**PURPLE·BLACK**

| DYE SOURCE | DYE RESULT | MORDANT | PROCESSING |
|---|---|---|---|
| ***Acer saccharinum*** SILVER MAPLE | | | |
| bark | dark rose-gray | iron | heat |
| | dark olive-gray | iron | decomposition |
| leaves | dark gray | iron | decomposition |
| ***Achillea millefolium*** YARROW | | | |
| leaves | dark olive-gray | iron | heat |
| ***Ageratina altissima*** WHITE SNAKEROOT | | | |
| roots | pale green-gray | copper | decomposition |
| ***Alnus maritima*** SEASIDE ALDER | | | |
| stems | medium khaki-gray | iron | heat |
| | medium gray | iron | decomposition |
| ***Ambrosia psilostachya*** WESTERN RAGWEED | | | |
| stems | light green-gray | iron | heat |
| ***Ambrosia trifida*** GIANT RAGWEED | | | |
| roots | light beige-gray | copper | decomposition |
| stems | medium olive-gray | iron | heat |
| ***Asclepias arenaria*** SAND MILKWEED | | | |
| leaves | light gray | iron | decomposition |
| ***Asimina triloba*** PAWPAW | | | |
| branches immature | medium khaki-gray | iron | heat |
| branches mature | medium gray | copper | decomposition |
| | light gray | iron | heat |
| | medium gray | none | decomposition |
| leaves | medium khaki-gray | iron | decomposition |
| rootstocks immature | medium khaki-gray | iron | heat |

| DYE SOURCE | DYE RESULT | MORDANT | PROCESSING |
|---|---|---|---|
| ***Baccharis salicina*** WILLOW BACCHARIS | | | |
| roots | medium khaki-gray | iron | heat |
| ***Baptisia australis*** BLUE WILD-INDIGO | | | |
| roots | light gray | copper | decomposition |
| | light gray | none | decomposition |
| ***Bidens bipinnata*** SPANISH NEEDLES | | | |
| leaves | brown-black | iron | heat |
| ***Brickellia eupatorioides*** FALSE BONESET | | | |
| roots | light olive-gray | iron | decomposition |
| stems | medium olive-gray | iron | heat |
| ***Buchnera americana*** BLUEHEARTS | | | |
| flowers | light gray | alum | heat |
| | light gray | none | heat |
| ***Bumelia lanuginosa*** CHITTAMWOOD | | | |
| fruits | light green-gray | alum | heat |
| ***Calylophus hartwegii*** HARTWEG'S SUNDROP | | | |
| roots | light gray | copper | decomposition |
| | light khaki-gray | iron | decomposition |
| | light gray | none | decomposition |
| ***Calylophus serrulatus*** HALFSHRUB SUNDROP | | | |
| flowers | light beige-gray | none | heat |
| leaves | brown-black | iron | heat |
| roots | light pink-gray | none | heat |
| ***Carya cordiformis*** BITTERNUT HICKORY | | | |
| leaves | medium khaki-gray | iron | heat |
| fruits | medium khaki-gray | iron | heat |
| ***Castanea ozarkensis*** OZARK CHINKAPIN | | | |
| branches | dark rose-gray | iron | heat |
| leaves | medium khaki-gray | iron | decomposition |
| | medium olive-gray | iron | heat |
| ***Catalpa speciosa*** NORTHERN CATALPA | | | |
| leaves | dark olive-gray | iron | heat |
| ***Ceanothus americanus*** NEW JERSEY TEA | | | |
| rhizomes and roots | medium pink-gray | iron | decomposition |
| stems | light beige-gray | iron | decomposition |
| ***Cephalanthus occidentalis*** BUTTONBUSH | | | |
| branches | medium gray | iron | heat |
| ***Cercis canadensis*** REDBUD | | | |
| bark of branches | dark gray | iron | heat |
| roots | medium gray | iron | heat |
| ***Chrysopsis pilosa*** SOFT GOLDEN-ASTER | | | |
| disk and ray florets | dark olive-gray | iron | heat |
| | pale green-gray | iron | decomposition |
| leaves | dark olive-gray | iron | heat |
| ***Cicuta maculata*** WATER-HEMLOCK | | | |
| roots | light olive-gray | copper | decomposition |

| DYE SOURCE | DYE RESULT | MORDANT | PROCESSING |
|---|---|---|---|
| ***Cirsium undulatum*** WAVY-LEAF THISTLE | | | |
| leaves | medium khaki-gray | copper | heat |
| | light beige-gray | none | heat |
| ***Coreopsis tinctoria*** PLAINS TICKSEED | | | |
| leaves | brown-black | iron | heat |
| ***Cornus drummondii*** ROUGH-LEAF DOGWOOD | | | |
| fruits | medium green-gray | iron | heat |
| pedicels | light green-gray | iron | heat |
| roots | dark olive-gray | iron | heat |
| | light green-gray | iron | decomposition |
| ***Cornus florida*** FLOWERING DOGWOOD | | | |
| bark | olive-gray | iron | heat |
| fruits | light olive-gray | iron | heat |
| | light pink-gray | none | heat |
| | pale green-gray | none | decomposition |
| leaves | brown-black | iron | decomposition |
| ***Cotinus obovatus*** SMOKE-TREE | | | |
| bark | light beige-gray | iron | decomposition |
| inflorescences | light khaki-gray | none | heat |
| leaves | medium olive-gray | iron | decomposition |
| ***Desmanthus illinoensis*** ILLINOIS BUNDLEFLOWER | | | |
| fruits | light gray | none | heat |
| leaves | brown-black | iron | heat |
| | light green-gray | iron | decomposition |
| stems | dark olive-gray | iron | heat |
| ***Diospyros virginiana*** PERSIMMON | | | |
| bark | light gray | iron | heat |
| ***Dracopis amplexicaulis*** CLASPING-LEAVED CONEFLOWER | | | |
| disk florets | medium gray | alum | heat |
| | medium khaki-gray | copper | heat |
| | light pink-gray | copper | decomposition |
| | dark teal-gray | iron | heat |
| | light pink-gray | iron | decomposition |
| | medium pink-gray | none | heat |
| leaves | light gray | none | decomposition |
| stems | light gray | iron | decomposition |
| ***Echinacea angustifolia*** PURPLE PRAIRIE CONEFLOWER | | | |
| leaves | olive-gray | iron | heat |
| ***Elephantopus carolinianus*** LEAFY ELEPHANT'S FOOT | | | |
| roots | light beige-gray | alum | decomposition |
| | light olive-gray | none | decomposition |
| stems | light green-gray | iron | heat |
| | light khaki-gray | none | decomposition |
| ***Euphorbia maculata*** SPOTTED SPURGE | | | |
| leaves and stems | brown-black | iron | heat |
| | light olive-gray | iron | decomposition |
| ***Fraxinus quadrangulata*** BLUE ASH | | | |
| bark | light beige-gray | iron | decomposition |

| DYE SOURCE | DYE RESULT | MORDANT | PROCESSING |
|---|---|---|---|
| ***Froelichia floridana*** FIELD SNAKE-COTTON | | | |
| leaves | medium khaki-gray | iron | decomposition |
| ***Gaillardia pulchella*** INDIAN BLANKET | | | |
| ray florets | light gray | alum | heat |
| | light gray | alum | decomposition |
| | light gray | copper | heat |
| | light gray | copper | decomposition |
| | medium green-gray | iron | heat |
| | light gray | iron | decomposition |
| | light gray | none | heat |
| ***Helianthus maximiliani*** MAXIMILIAN'S SUNFLOWER | | | |
| rhizomes | medium olive-gray | iron | heat |
| stems | medium olive-gray | iron | heat |
| ***Helianthus mollis*** ASHY SUNFLOWER | | | |
| disk florets | black | iron | heat |
| leaves | brown-black | iron | heat |
| ***Helianthus tuberosus*** JERUSALEM ARTICHOKE | | | |
| leaves | light beige-gray | none | decomposition |
| roots | light beige-gray | copper | decomposition |
| stems | light gray | none | decomposition |
| ***Hydrangea arborescens*** WILD HYDRANGEA | | | |
| leaves | light gray | iron | decomposition |
| ***Ipomoea leptophylla*** BUSH MORNING-GLORY | | | |
| rootstocks | medium khaki-gray | iron | heat |
| | medium khaki-gray | iron | decomposition |
| ***Juniperus virginiana*** EASTERN RED CEDAR | | | |
| bark | medium gray | iron | heat |
| roots | light blue-gray | alum | decomposition |
| | medium blue-gray | copper | decomposition |
| | light gray | iron | heat |
| | light green-gray | iron | decomposition |
| | light gray | tin | decomposition |
| | light blue-gray | none | decomposition |
| ***Lesquerella gracilis*** SPREADING BLADDERPOD | | | |
| leaves and stems | light khaki-gray | iron | decomposition |
| | light olive-gray | none | decomposition |
| ***Liatris punctata*** DOTTED GAYFEATHER | | | |
| phyllaries | light beige-gray | none | decomposition |
| ***Liquidambar styraciflua*** SWEET GUM | | | |
| bark of branches | purple-black | iron | heat |
| bark of trunk | black | iron | decomposition |
| roots | olive-gray | iron | heat |
| ***Maclura pomifera*** OSAGE ORANGE | | | |
| fruits | dark olive-gray | iron | heat |
| ***Mentzelia nuda*** BRACTLESS BLAZING-STAR | | | |
| leaves | light beige-gray | iron | heat |
| ***Neptunia lutea*** YELLOW NEPTUNE | | | |
| flower buds | dark olive-gray | iron | heat |

| DYE SOURCE | DYE RESULT | MORDANT | PROCESSING |
|---|---|---|---|
| ***Nyssa sylvatica*** BLACK GUM | | | |
| bark | medium gray | iron | heat |
| branchlets | light olive-gray | iron | heat |
| leaves | medium gray | iron | heat |
| ***Oenothera heterophylla*** SAND EVENING-PRIMROSE | | | |
| flowers | dark gray | iron | heat |
| roots | dark gray | iron | heat |
| stems | dark gray | iron | heat |
| | dark gray | iron | decomposition |
| ***Oenothera macrocarpa*** BIGFRUIT EVENING-PRIMROSE | | | |
| fruits immature | dark gray | iron | decomposition |
| roots | medium khaki-gray | iron | heat |
| stems and bracts | light gray | copper | heat |
| | medium olive-gray | iron | heat |
| | light gray | none | heat |
| ***Parthenocissus quinquefolia*** VIRGINIA CREEPER | | | |
| bark | medium olive-gray | iron | heat |
| | light gray | iron | decomposition |
| fruits | medium khaki-gray | iron | decomposition |
| ***Phlox pilosa*** PRAIRIE PHLOX | | | |
| stems | light beige-gray | copper | decomposition |
| ***Podophyllum peltatum*** MAY-APPLE | | | |
| leaves | light gray | iron | decomposition |
| ***Polygonum pensylvanicum*** PENNSYLVANIA SMARTWEED | | | |
| flowers | olive-black | iron | heat |
| leaves | pale green-gray | iron | decomposition |
| stems | light khaki-gray | copper | heat |
| ***Populus deltoides*** EASTERN COTTONWOOD | | | |
| roots | medium khaki-gray | iron | heat |
| ***Potentilla arguta*** TALL CINQUEFOIL | | | |
| leaves | dark olive-gray | iron | heat |
| rootstocks | light gray | copper | decomposition |
| | medium gray | iron | decomposition |
| ***Ratibida columnifera*** MEXICAN HAT | | | |
| disk florets | light green-gray | copper | heat |
| | light gray | none | heat |
| ***Rhus copallinum*** WINGED SUMAC | | | |
| bark | medium gray | iron | decomposition |
| flowers | dark gray | iron | decomposition |
| ***Rhus glabra*** SMOOTH SUMAC | | | |
| fruits | medium gray | iron | decomposition |
| leaves | medium olive-gray | iron | decomposition |
| ***Robinia pseudoacacia*** BLACK LOCUST | | | |
| leaves | light beige-gray | copper | decomposition |
| | dark olive-gray | iron | heat |
| | light khaki-gray | iron | decomposition |
| ***Rudbeckia grandiflora*** ROUGH CONEFLOWER | | | |
| disk florets | medium olive-gray | iron | heat |
| | medium khaki-gray | iron | decomposition |

| DYE SOURCE | DYE RESULT | MORDANT | PROCESSING |
|---|---|---|---|
| ***Rudbeckia hirta*** BLACK-EYED SUSAN | | | |
| disk florets | medium green-gray | alum | heat |
| | light gray | copper | decomposition |
| | light gray | iron | decomposition |
| | medium gray | none | heat |
| leaves | black | iron | heat |
| ***Salix exigua*** SANDBAR WILLOW | | | |
| stems | light khaki-gray | iron | heat |
| | pink-gray | none | decomposition |
| ***Salix nigra*** BLACK WILLOW | | | |
| bark | light gray | iron | heat |
| ***Sassafras albidum*** SASSAFRAS | | | |
| roots | light gray | iron | heat |
| ***Silphium integrifolium*** PRAIRIE ROSINWEED | | | |
| roots | light beige-gray | none | decomposition |
| ***Solanum dimidiatum*** WESTERN HORSE-NETTLE | | | |
| roots | light gray | none | decomposition |
| ***Solanum elaeagnifolium*** SILVERLEAF NIGHTSHADE | | | |
| leaves | green-gray | copper | decomposition |
| ***Solidago rigida*** STIFF GOLDENROD | | | |
| heads immature | olive-black | iron | heat |
| heads mature | dark olive-gray | iron | heat |
| leaves | dark olive-gray | iron | heat |
| roots | light green-gray | copper | decomposition |
| ***Stillingia sylvatica*** QUEEN'S DELIGHT | | | |
| leaves | dark olive-gray | iron | decomposition |
| roots | medium gray | copper | decomposition |
| ***Teucrium canadense*** AMERICAN GERMANDER | | | |
| leaves | medium gray | iron | decomposition |
| | light gray | none | decomposition |
| ***Vaccinium arboreum*** FARKLEBERRY | | | |
| leaves | medium gray | alum | heat |
| | medium gray | copper | heat |
| | medium gray | none | heat |
| ***Vernonia baldwinii*** WESTERN IRONWEED | | | |
| flower buds | light green-gray | alum | heat |
| leaves | dark olive-gray | iron | heat |
| stems | dark olive-gray | iron | heat |
| ***Vitis aestivalis*** PIGEON GRAPE | | | |
| bark | light gray | iron | heat |
| fruits | pink-gray | none | heat |

# Materials That Produce Little or No Color

The unbleached wool yarns frequently used for natural dyeing are usually cream or beige to begin with. If these yarns remain cream or beige after undergoing the dyeing process, it can be concluded that the vegetative material used as a dyestuff possesses very little dye potential or effect. This is true of the natural dye materials listed in this chapter, all of which have limited usefulness as colorants. Often this limitation is restricted to a specific part of a particular plant species, or to a particular mordant, or to a method of dye processing. Even if the roots of a plant produce very little color, the leaves of the same plant may be a rich source of color. Likewise, even if the roots produce color when heated, the same roots may produce little color when decomposed. For this reason, many of the plants listed here were also listed in previous chapters as producers of natural dye. It is important to note which plant organs, mordants, or processing methods tend to produce disappointing dye results and not totally discount an entire plant as a potential dyestuff.

Our search for natural dyes resulted in various beige colors, shown here with Munsell notations. Following these color chips is a list of the plants that were used to obtain these colors, including mordant and processing conditions.

5Y 8.5/2 7.5Y 8.5/2

**BEIGE**

2.5Y 8.5/2 7.5Y 8/2 10Y 8/2 10YR 8/2

**GRAY·BEIGE**

5Y 8/4

**KHAKI·BEIGE**

10Y 8.5/2

**LIGHT GRAY·BEIGE**

10Y 8/4

**LIGHT OLIVE·BEIGE**

2.5Y 9/2

**TAN·BEIGE**

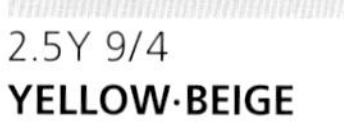

2.5Y 9/4 5Y 8.5/4

**YELLOW·BEIGE**

| DYE SOURCE | DYE RESULT | MORDANT | PROCESSING |
|---|---|---|---|
| ***Acer negundo*** BOXELDER | | | |
| leaves | yellow-beige | copper | heat |
| | khaki-beige | iron | heat |
| | yellow-beige | iron | decomposition |
| ***Acer saccharinum*** SILVER MAPLE | | | |
| bark | yellow-beige | none | heat |
| ***Achillea millefolium*** YARROW | | | |
| flowers | light olive-beige | iron | decomposition |
| ***Ageratina altissima*** WHITE SNAKEROOT | | | |
| heads | yellow-beige | copper | decomposition |
| | yellow-beige | iron | decomposition |
| leaves | khaki-beige | alum | decomposition |
| | yellow-beige | none | decomposition |
| roots | beige | alum | decomposition |
| | beige | iron | heat |
| | gray-beige | none | decomposition |
| ***Alnus maritima*** SEASIDE ALDER | | | |
| leaves | khaki-beige | none | heat |
| | yellow-beige | none | decomposition |
| stems | yellow-beige | tin | decomposition |
| | beige | none | heat |
| ***Ambrosia psilostachya*** WESTERN RAGWEED | | | |
| leaves and heads | yellow-beige | alum | decomposition |
| rhizomes and roots | gray-beige | iron | heat |
| | khaki-beige | iron | decomposition |
| ***Ambrosia trifida*** GIANT RAGWEED | | | |
| roots | beige | alum | decomposition |
| | khaki-beige | copper | heat |
| | beige | iron | decomposition |
| stems | light olive-beige | none | heat |
| ***Amorpha canescens*** LEADPLANT | | | |
| bark | yellow-beige | alum | decomposition |
| | yellow-beige | copper | decomposition |
| | yellow-beige | iron | decomposition |
| | yellow-beige | none | heat |
| leaves | light olive-beige | iron | decomposition |
| spikes | khaki-beige | alum | heat |
| | khaki-beige | none | heat |
| ***Apocynum cannabinum*** INDIAN HEMP | | | |
| roots | khaki-beige | copper | decomposition |
| | yellow-beige | iron | decomposition |
| | tan-beige | none | heat |
| stems | gray-beige | alum | decomposition |
| | gray-beige | none | decomposition |
| ***Argemone polyanthemos*** PRICKLY POPPY | | | |
| leaves | khaki-beige | iron | decomposition |
| pistils | yellow-beige | copper | heat |
| | khaki-beige | iron | heat |
| | khaki-beige | iron | decomposition |

| DYE SOURCE | DYE RESULT | MORDANT | PROCESSING |
|---|---|---|---|
| ***Argemone polyanthemos,*** continued | | | |
| roots | gray-beige | iron | decomposition |
| | yellow-beige | tin | decomposition |
| | gray-beige | none | decomposition |
| ***Asclepias arenaria*** SAND MILKWEED | | | |
| flowers | khaki-beige | iron | heat |
| leaves | gray-beige | alum | decomposition |
| | gray-beige | none | decomposition |
| ***Asclepias tuberosa*** BUTTERFLY MILKWEED | | | |
| flowers | yellow-beige | alum | heat |
| leaves | yellow-beige | copper | decomposition |
| stems | khaki-beige | copper | heat |
| ***Asclepias viridis*** ANTELOPE-HORN MILKWEED | | | |
| flowers | yellow-beige | iron | heat |
| roots | yellow-beige | copper | heat |
| | yellow-beige | iron | heat |
| ***Asimina triloba*** PAWPAW | | | |
| branches immature | beige | copper | heat |
| | gray-beige | none | heat |
| branches mature | tan-beige | alum | heat |
| | gray-beige | copper | heat |
| | khaki-beige | tin | decomposition |
| | tan-beige | none | heat |
| leaves immature | yellow-beige | alum | heat |
| | yellow-beige | alum | decomposition |
| | yellow-beige | copper | decomposition |
| | yellow-beige | tin | decomposition |
| | yellow-beige | none | decomposition |
| leaves mature | gray-beige | none | decomposition |
| rootstocks immature | yellow-beige | alum | heat |
| ***Baccharis salicina*** WILLOW BACCHARIS | | | |
| leaves | khaki-beige | copper | heat |
| roots | yellow-beige | alum | heat |
| | beige | copper | decomposition |
| | gray-beige | iron | decomposition |
| | beige | none | decomposition |
| stems | khaki-beige | alum | decomposition |
| ***Baptisia australis*** BLUE WILD-INDIGO | | | |
| petals | khaki-beige | alum | heat |
| roots | gray-beige | alum | decomposition |
| stems (basal) | yellow-beige | alum | heat |
| ***Baptisia bracteata*** PLAINS WILD-INDIGO | | | |
| leaves and stems | gray-beige | none | decomposition |
| roots | yellow-beige | alum | decomposition |
| | khaki-beige | copper | decomposition |
| | beige | none | decomposition |
| ***Baptisia sphaerocarpa*** GOLDEN WILD-INDIGO | | | |
| fruits and bracts | yellow-beige | alum | decomposition |
| | yellow-beige | copper | decomposition |

| DYE SOURCE | DYE RESULT | MORDANT | PROCESSING |
|---|---|---|---|
| | khaki-beige | iron | decomposition |
| leaves | khaki-beige | alum | decomposition |
| | yellow-beige | copper | heat |
| | yellow-beige | tin | decomposition |
| | gray-beige | none | decomposition |
| petals | yellow-beige | copper | heat |
| roots | khaki-beige | copper | decomposition |
| | yellow-beige | iron | decomposition |
| | yellow-beige | none | heat |
| stems | yellow-beige | alum | heat |
| | khaki-beige | copper | decomposition |
| | yellow-beige | iron | heat |
| | khaki-beige | iron | decomposition |
| | tan-beige | none | decomposition |
| ***Bidens bipinnata*** SPANISH NEEDLES | | | |
| roots | khaki-beige | copper | decomposition |
| ***Brickellia eupatorioides*** FALSE BONESET | | | |
| roots | beige | alum | decomposition |
| | light olive-beige | copper | heat |
| | light olive-beige | iron | decomposition |
| | light gray-beige | none | decomposition |
| stems | khaki-beige | copper | decomposition |
| ***Buchnera americana*** BLUEHEARTS | | | |
| leaves | light olive-beige | none | heat |
| ***Bumelia lanuginosa*** CHITTAMWOOD | | | |
| branchlets | khaki-beige | copper | heat |
| | yellow-beige | iron | heat |
| | yellow-beige | iron | decomposition |
| fruits | yellow-beige | copper | decomposition |
| | light gray-beige | none | heat |
| leaves | light olive-beige | iron | heat |
| ***Callicarpa americana*** AMERICAN BEAUTYBERRY | | | |
| fruits | light olive-beige | iron | decomposition |
| | yellow-beige | none | decomposition |
| ***Calylophus hartwegii*** HARTWEG'S SUNDROP | | | |
| petals | tan-beige | none | heat |
| ***Calylophus serrulatus*** HALFSHRUB SUNDROP | | | |
| leaves | khaki-beige | none | decomposition |
| stems | yellow-beige | copper | decomposition |
| | yellow-beige | iron | decomposition |
| | gray-beige | none | heat |
| ***Campsis radicans*** TRUMPET CREEPER | | | |
| leaves | gray-beige | none | decomposition |
| ***Carya cordiformis*** BITTERNUT HICKORY | | | |
| leaves | yellow-beige | alum | decomposition |
| | yellow-beige | none | decomposition |
| ***Carya illinoinensis*** PECAN | | | |
| branchlets | khaki-beige | alum | decomposition |
| | khaki-beige | none | decomposition |

| DYE SOURCE | DYE RESULT | MORDANT | PROCESSING |
|---|---|---|---|
| ***Castanea ozarkensis*** OZARK CHINKAPIN | | | |
| leaves | yellow-beige | copper | decomposition |
| | beige | none | decomposition |
| ***Castilleja indivisa*** INDIAN PAINTBRUSH | | | |
| bract tips | yellow-beige | alum | heat |
| | yellow-beige | alum | decomposition |
| | yellow-beige | copper | decomposition |
| ***Castilleja purpurea*** var. ***citrina*** YELLOW PAINTBRUSH | | | |
| leaves | yellow-beige | alum | heat |
| | khaki-beige | copper | heat |
| | tan-beige | none | decomposition |
| petals and bracts | yellow-beige | none | decomposition |
| ***Ceanothus americanus*** NEW JERSEY TEA | | | |
| flowers | yellow-beige | copper | decomposition |
| | yellow-beige | iron | decomposition |
| leaves | yellow-beige | copper | decomposition |
| rhizomes and roots | tan-beige | alum | decomposition |
| | tan-beige | none | decomposition |
| stems | tan-beige | none | decomposition |
| ***Cephalanthus occidentalis*** BUTTONBUSH | | | |
| flowers | khaki-beige | copper | heat |
| | khaki-beige | iron | decomposition |
| leaves | yellow-beige | copper | heat |
| | khaki-beige | iron | heat |
| | yellow-beige | iron | decomposition |
| ***Cercis canadensis*** REDBUD | | | |
| flowers | yellow-beige | alum | heat |
| | khaki-beige | copper | heat |
| ***Chrysopsis pilosa*** SOFT GOLDEN-ASTER | | | |
| leaves | khaki-beige | copper | decomposition |
| | yellow-beige | iron | decomposition |
| ***Cicuta maculata*** WATER-HEMLOCK | | | |
| leaves | khaki-beige | copper | decomposition |
| roots | gray-beige | alum | decomposition |
| | gray-beige | iron | decomposition |
| | light gray-beige | none | decomposition |
| stems | yellow-beige | iron | heat |
| ***Cirsium undulatum*** WAVY-LEAF THISTLE | | | |
| leaves | khaki-beige | alum | heat |
| | yellow-beige | alum | decomposition |
| | yellow-beige | copper | decomposition |
| | khaki-beige | iron | decomposition |
| | beige | none | decomposition |
| roots | tan-beige | none | heat |
| ***Comandra umbellata*** BASTARD TOAD-FLAX | | | |
| leaves | yellow-beige | none | heat |
| stems | tan-beige | none | heat |

| DYE SOURCE | DYE RESULT | MORDANT | PROCESSING |
|---|---|---|---|
| ***Cornus drummondii*** ROUGH-LEAF DOGWOOD | | | |
| branchlets | yellow-beige | tin | decomposition |
| | beige | none | decomposition |
| pedicels | light olive-beige | copper | heat |
| | yellow-beige | copper | decomposition |
| ***Cornus florida*** FLOWERING DOGWOOD | | | |
| fruits | yellow-beige | alum | decomposition |
| | yellow-beige | copper | heat |
| ***Cotinus obovatus*** SMOKE-TREE | | | |
| bark | yellow-beige | copper | decomposition |
| | yellow-beige | tin | decomposition |
| leaves | yellow-beige | none | heat |
| ***Cucurbita foetidissima*** BUFFALO GOURD | | | |
| leaves | yellow-beige | alum | heat |
| stems | yellow-beige | alum | decomposition |
| | khaki-beige | copper | heat |
| | yellow-beige | iron | heat |
| | yellow-beige | iron | decomposition |
| | yellow-beige | tin | decomposition |
| ***Cycloloma atriplicifolium*** TUMBLE RINGWING | | | |
| leaves and inflorescences | beige | copper | decomposition |
| ***Cyperus squarrosus*** BEARDED FLATSEDGE | | | |
| leaves | khaki-beige | copper | decomposition |
| | khaki-beige | iron | decomposition |
| | yellow-beige | none | heat |
| | tan-beige | none | decomposition |
| spikes | yellow-beige | alum | heat |
| | khaki-beige | iron | decomposition |
| | yellow-beige | none | heat |
| | tan-beige | none | decomposition |
| ***Dalea candida*** WHITE PRAIRIE-CLOVER | | | |
| flowers | khaki-beige | copper | heat |
| roots | yellow-beige | alum | decomposition |
| | khaki-beige | copper | decomposition |
| | yellow-beige | none | decomposition |
| ***Dalea lanata*** WOOLLY DALEA | | | |
| leaves | khaki-beige | tin | heat |
| ***Dalea purpurea*** PURPLE PRAIRIE-CLOVER | | | |
| flowers | yellow-beige | copper | heat |
| leaves and stems | light gray-beige | none | decomposition |
| ***Dalea villosa*** SILKY PRAIRIE-CLOVER | | | |
| flowers | khaki-beige | copper | decomposition |
| roots | yellow-beige | copper | decomposition |
| | yellow-beige | iron | decomposition |
| stems | yellow-beige | copper | heat |
| | light gray-beige | copper | decomposition |
| | khaki-beige | iron | decomposition |

| DYE SOURCE | DYE RESULT | MORDANT | PROCESSING |
|---|---|---|---|
| ***Desmanthus illinoensis*** ILLINOIS BUNDLEFLOWER | | | |
| fruits | khaki-beige | copper | decomposition |
| | yellow-beige | iron | decomposition |
| leaves | gray-beige | copper | decomposition |
| | khaki-beige | none | heat |
| stems | yellow-beige | tin | decomposition |
| ***Desmodium glutinosum*** STICKY TICKCLOVER | | | |
| leaves | khaki-beige | copper | heat |
| | yellow-beige | copper | decomposition |
| | yellow-beige | iron | decomposition |
| stems | tan-beige | alum | heat |
| | beige | copper | heat |
| | beige | iron | heat |
| | tan-beige | none | heat |
| | gray-beige | none | decomposition |
| ***Diospyros virginiana*** PERSIMMON | | | |
| bark | khaki-beige | copper | decomposition |
| branchlets | yellow-beige | alum | decomposition |
| | khaki-beige | iron | decomposition |
| leaves | khaki-beige | copper | decomposition |
| | khaki-beige | iron | decomposition |
| | yellow-beige | tin | decomposition |
| fruit immature | yellow-beige | copper | decomposition |
| | yellow-beige | iron | heat |
| fruit mature | light olive-beige | copper | heat |
| | gray-beige | iron | heat |
| | tan-beige | none | heat |
| ***Dracopis amplexicaulis*** CLASPING-LEAVED CONEFLOWER | | | |
| ray florets | khaki-beige | alum | heat |
| | khaki-beige | copper | heat |
| | khaki-beige | iron | decomposition |
| | tan-beige | none | decomposition |
| | beige | none | heat |
| roots | khaki-beige | alum | heat |
| | light gray-beige | copper | decomposition |
| | beige | none | heat |
| stems | tan-beige | alum | decomposition |
| | khaki-beige | copper | heat |
| ***Echinacea angustifolia*** PURPLE PRAIRIE CONEFLOWER | | | |
| disk florets | khaki-beige | alum | heat |
| | khaki-beige | tin | decomposition |
| | yellow-beige | none | heat |
| | gray-beige | none | decomposition |
| leaves | khaki-beige | none | heat |
| | khaki-beige | none | decomposition |
| roots | khaki-beige | copper | heat |
| ***Echinocereus reichenbachii*** LACE HEDGEHOG-CACTUS | | | |
| flowers | yellow-beige | copper | heat |
| | yellow-beige | copper | decomposition |
| | beige | none | decomposition |

| DYE SOURCE | DYE RESULT | MORDANT | PROCESSING |
|---|---|---|---|
| ***Elephantopus carolinianus*** LEAFY ELEPHANT'S FOOT | | | |
| leaves | yellow-beige | alum | decomposition |
| | yellow-beige | copper | heat |
| | light gray-beige | none | decomposition |
| stems | yellow-beige | tin | decomposition |
| ***Euphorbia maculata*** SPOTTED SPURGE | | | |
| leaves and stems | gray-beige | copper | decomposition |
| ***Euphorbia marginata*** SNOW-ON-THE-MOUNTAIN | | | |
| bracts | khaki-beige | copper | decomposition |
| | yellow-beige | iron | decomposition |
| | yellow-beige | tin | heat |
| cyathia | khaki-beige | copper | decomposition |
| leaves | yellow-beige | alum | decomposition |
| | khaki-beige | copper | decomposition |
| | yellow-beige | iron | decomposition |
| roots | yellow-beige | alum | heat |
| | yellow-beige | copper | decomposition |
| | yellow-beige | iron | decomposition |
| | beige | none | decomposition |
| stems | yellow-beige | copper | decomposition |
| | khaki-beige | iron | decomposition |
| ***Eustoma exaltatum*** PRAIRIE GENTIAN | | | |
| petals | yellow-beige | copper | decomposition |
| | khaki-beige | iron | decomposition |
| ***Fraxinus pennsylvanica*** GREEN ASH | | | |
| bark | yellow-beige | alum | heat |
| fruits | yellow-beige | copper | decomposition |
| | yellow-beige | iron | decomposition |
| leaves | yellow-beige | alum | decomposition |
| | yellow-beige | none | heat |
| | yellow-beige | none | decomposition |
| ***Fraxinus quadrangulata*** BLUE ASH | | | |
| bark | yellow-beige | none | heat |
| leaves | yellow-beige | copper | decomposition |
| | yellow-beige | none | decomposition |
| ***Froelichia floridana*** FIELD SNAKE-COTTON | | | |
| flowers | beige | iron | decomposition |
| leaves | gray-beige | copper | decomposition |
| | yellow-beige | tin | decomposition |
| | beige | none | decomposition |
| ***Gaillardia aestivalis*** PRAIRIE GAILLARDIA | | | |
| disk florets | yellow-beige | none | heat |
| leaves | light olive-beige | copper | decomposition |
| | yellow-beige | none | decomposition |
| roots | yellow-beige | copper | heat |
| | khaki-beige | iron | heat |
| ***Gaillardia pulchella*** INDIAN BLANKET | | | |
| disk florets | yellow-beige | alum | heat |
| | khaki-beige | copper | heat |
| | yellow-beige | iron | decomposition |

| DYE SOURCE | DYE RESULT | MORDANT | PROCESSING |
|---|---|---|---|
| ***Gaillardia pulchella,*** continued | | | |
| roots | beige | copper | decomposition |
| | gray-beige | iron | decomposition |
| | gray-beige | none | decomposition |
| ***Gleditsia triacanthos*** HONEY LOCUST | | | |
| fruits | yellow-beige | iron | heat |
| ***Gymnocladus dioicus*** KENTUCKY COFFEE-TREE | | | |
| bark | khaki-beige | iron | decomposition |
| fruits | yellow-beige | copper | decomposition |
| | yellow-beige | iron | decomposition |
| petioles | yellow-beige | copper | decomposition |
| seeds | yellow-beige | copper | heat |
| | yellow-beige | copper | decomposition |
| stems | yellow-beige | alum | decomposition |
| | yellow-beige | copper | heat |
| | yellow-beige | copper | decomposition |
| | yellow-beige | iron | heat |
| | yellow-beige | iron | decomposition |
| | yellow-beige | none | heat |
| ***Hedyotis nigricans*** PRAIRIE BLUETS | | | |
| roots | tan-beige | none | heat |
| ***Helenium amarum*** BITTER SNEEZEWEED | | | |
| roots | yellow-beige | copper | decomposition |
| stems | yellow-beige | copper | heat |
| ***Helianthus annuus*** ANNUAL SUNFLOWER | | | |
| disk florets | yellow-beige | alum | decomposition |
| | yellow-beige | copper | decomposition |
| | tan-beige | none | decomposition |
| leaves | yellow-beige | alum | decomposition |
| | light olive-beige | none | heat |
| | yellow-beige | none | decomposition |
| ray florets | yellow-beige | copper | heat |
| ***Helianthus maximiliani*** MAXIMILIAN'S SUNFLOWER | | | |
| leaves | beige | none | decomposition |
| stems | yellow-beige | alum | decomposition |
| | beige | none | decomposition |
| ***Helianthus mollis*** ASHY SUNFLOWER | | | |
| ray florets | yellow-beige | none | heat |
| ***Helianthus tuberosus*** JERUSALEM ARTICHOKE | | | |
| disk and ray florets | yellow-beige | iron | decomposition |
| leaves | beige | alum | decomposition |
| | yellow-beige | tin | decomposition |
| roots | yellow-beige | iron | decomposition |
| | beige | none | decomposition |
| stems | light olive-beige | copper | heat |
| ***Hibiscus laevis*** SCARLET ROSE MALLOW | | | |
| leaves | tan-beige | alum | decomposition |
| | tan-beige | none | decomposition |

| DYE SOURCE | DYE RESULT | MORDANT | PROCESSING |
|---|---|---|---|
| petals | yellow-beige | copper | decomposition |
| sepals | beige | copper | decomposition |
| | yellow-beige | iron | heat |
| | beige | iron | decomposition |
| stems | khaki-beige | iron | heat |
| ***Hydrangea arborescens*** WILD HYDRANGEA | | | |
| branches | yellow-beige | copper | heat |
| | yellow-beige | iron | heat |
| flowers | yellow-beige | copper | decomposition |
| | khaki-beige | iron | decomposition |
| leaves | khaki-beige | copper | heat |
| ***Ipomoea leptophylla*** BUSH MORNING-GLORY | | | |
| leaves | yellow-beige | none | heat |
| rootstocks | khaki-beige | copper | heat |
| | yellow-beige | tin | decomposition |
| stems | yellow-beige | alum | heat |
| | yellow-beige | copper | heat |
| ***Ipomopsis rubra*** STANDING CYPRESS | | | |
| petals | yellow-beige | alum | heat |
| | yellow-beige | copper | heat |
| roots | yellow-beige | copper | heat |
| stems | yellow-beige | copper | heat |
| ***Juniperus virginiana*** EASTERN RED CEDAR | | | |
| bark | yellow-beige | copper | decomposition |
| ***Lesquerella gracilis*** SPREADING BLADDERPOD | | | |
| stems and leaves | gray-beige | alum | decomposition |
| | khaki-beige | copper | heat |
| | khaki-beige | tin | decomposition |
| ***Liatris punctata*** DOTTED GAYFEATHER | | | |
| florets | yellow-beige | iron | heat |
| leaves and stems | light olive-beige | alum | decomposition |
| | yellow-beige | none | decomposition |
| phyllaries | yellow-beige | alum | heat |
| | yellow-beige | alum | decomposition |
| | yellow-beige | copper | heat |
| roots | yellow-beige | iron | decomposition |
| | yellow-beige | tin | heat |
| ***Liatris squarrosa*** SCALY GAYFEATHER | | | |
| flowers | beige | alum | heat |
| leaves | yellow-beige | copper | decomposition |
| | yellow-beige | iron | decomposition |
| | yellow-beige | none | decomposition |
| roots | khaki-beige | copper | heat |
| | yellow-beige | iron | heat |
| ***Linum sulcatum*** GROOVED FLAX | | | |
| leaves, stems, and petals | yellow-beige | tin | heat |
| | yellow-beige | none | heat |
| roots | yellow-beige | tin | heat |

| DYE SOURCE | DYE RESULT | MORDANT | PROCESSING |
|---|---|---|---|
| ***Maclura pomifera*** OSAGE ORANGE | | | |
| bark | yellow-beige | copper | decomposition |
| | khaki-beige | iron | decomposition |
| leaves | yellow-beige | alum | decomposition |
| | khaki-beige | copper | decomposition |
| | yellow-beige | none | heat |
| | yellow-beige | none | decomposition |
| roots | light olive-beige | iron | decomposition |
| ***Mentzelia nuda*** BRACTLESS BLAZING-STAR | | | |
| leaves | beige | copper | decomposition |
| | yellow-beige | iron | decomposition |
| | yellow-beige | tin | heat |
| stems | gray-beige | alum | heat |
| | beige | copper | decomposition |
| | gray-beige | none | heat |
| ***Monarda fistulosa*** WILD BERGAMOT | | | |
| flowers | light olive-beige | iron | decomposition |
| leaves | yellow-beige | copper | decomposition |
| ***Morus rubra*** RED MULBERRY | | | |
| bark | yellow-beige | copper | heat |
| | yellow-beige | copper | decomposition |
| | yellow-beige | iron | heat |
| | yellow-beige | iron | decomposition |
| bark green | yellow-beige | copper | decomposition |
| branches | yellow-beige | copper | decomposition |
| branchlets | yellow-beige | copper | decomposition |
| leaves | khaki-beige | copper | decomposition |
| | yellow-beige | iron | heat |
| ***Neptunia lutea*** YELLOW NEPTUNE | | | |
| leaves | light gray-beige | none | decomposition |
| | yellow-beige | none | heat |
| ***Nuttallanthus canadensis*** TOAD-FLAX | | | |
| leaves and stems | yellow-beige | iron | decomposition |
| ***Nyssa sylvatica*** BLACK GUM | | | |
| bark | khaki-beige | alum | decomposition |
| | khaki-beige | iron | decomposition |
| branchlets | yellow-beige | copper | heat |
| | khaki-beige | iron | decomposition |
| ***Oenothera heterophylla*** SAND EVENING-PRIMROSE | | | |
| leaves | yellow-beige | none | decomposition |
| stems | yellow-beige | alum | heat |
| | yellow-beige | none | decomposition |
| ***Oenothera macrocarpa*** BIGFRUIT EVENING-PRIMROSE | | | |
| bracts and stems | beige | alum | heat |
| | khaki-beige | alum | decomposition |
| | yellow-beige | copper | decomposition |
| fruits mature | khaki-beige | copper | heat |
| fruits immature | tan-beige | none | heat |
| leaves | yellow-beige | alum | heat |

| DYE SOURCE | DYE RESULT | MORDANT | PROCESSING |
|---|---|---|---|
| petals | yellow-beige | alum | heat |
| | khaki-beige | iron | decomposition |
| | yellow-beige | none | heat |
| roots | gray-beige | none | heat |
| stems | khaki-beige | alum | heat |
| | gray-beige | none | heat |
| ***Opuntia macrorhiza*** PRICKLY-PEAR | | | |
| fruits | beige | alum | decomposition |
| | yellow-beige | copper | heat |
| | beige | copper | decomposition |
| | yellow-beige | iron | heat |
| | beige | iron | decomposition |
| | beige | none | decomposition |
| ***Packera obovata*** ROUNDLEAF GROUNDSEL | | | |
| flowers | yellow-beige | copper | heat |
| leaves and stems | yellow-beige | none | decomposition |
| roots | yellow-beige | tin | heat |
| | yellow-beige | none | heat |
| ***Parthenocissus quinquefolia*** VIRGINIA CREEPER | | | |
| bark | tan-beige | alum | decomposition |
| | tan-beige | tin | decomposition |
| | yellow-beige | none | decomposition |
| fruits | yellow-beige | copper | heat |
| pedicels | yellow-beige | copper | decomposition |
| | gray-beige | iron | heat |
| | yellow-beige | iron | decomposition |
| sepals | yellow-beige | copper | heat |
| ***Pediomelum cuspidatum*** TALL-BREAD SCURFPEA | | | |
| fruits | yellow-beige | copper | heat |
| | khaki-beige | copper | decomposition |
| | khaki-beige | iron | decomposition |
| leaves | khaki-beige | iron | decomposition |
| | yellow-beige | none | decomposition |
| ***Phlox pilosa*** PRAIRIE PHLOX | | | |
| leaves | yellow-beige | iron | decomposition |
| stems | yellow-beige | copper | heat |
| | yellow-beige | iron | heat |
| ***Phyla lanceolata*** LANCELEAF FROG-FRUIT | | | |
| leaves | yellow-beige | copper | decomposition |
| | khaki-beige | iron | decomposition |
| stems | yellow-beige | copper | heat |
| | yellow-beige | copper | decomposition |
| | khaki-beige | iron | heat |
| ***Physalis angulata*** CUTLEAF GROUND-CHERRY | | | |
| fruits | beige | iron | decomposition |
| leaves | yellow-beige | copper | heat |
| | yellow-beige | copper | decomposition |
| | khaki-beige | iron | heat |
| | khaki-beige | iron | decomposition |
| | yellow-beige | none | decomposition |

| DYE SOURCE | DYE RESULT | MORDANT | PROCESSING |
|---|---|---|---|
| ***Physalis virginiana*** VIRGINIA GROUND-CHERRY | | | |
| leaves | light olive-beige | none | heat |
| ***Phytolacca americana*** POKEWEED | | | |
| leaves | yellow-beige | copper | heat |
| | yellow-beige | iron | heat |
| stems | yellow-beige | tin | heat |
| ***Platanus occidentalis*** SYCAMORE | | | |
| bark | yellow-beige | alum | decomposition |
| | yellow-beige | none | decomposition |
| ***Podophyllum peltatum*** MAY-APPLE | | | |
| rhizomes and roots | khaki-beige | copper | decomposition |
| | yellow-beige | none | heat |
| | tan-beige | none | decomposition |
| ***Polygala alba*** WHITE MILKWORT | | | |
| flowers | yellow-beige | copper | decomposition |
| | khaki-beige | iron | decomposition |
| ***Polygonum pensylvanicum*** PENNSYLVANIA SMARTWEED | | | |
| leaves | yellow-beige | tin | decomposition |
| | gray-beige | none | heat |
| | light gray-beige | none | decomposition |
| stems | khaki-beige | copper | decomposition |
| | light gray-beige | none | decomposition |
| ***Polygonum punctatum*** DOTTED SMARTWEED | | | |
| flowers and flower buds | khaki-beige | iron | decomposition |
| | light olive-beige | none | heat |
| leaves | light olive-beige | none | heat |
| | light olive-beige | none | decomposition |
| stems | light olive-beige | copper | heat |
| | yellow-beige | copper | decomposition |
| | yellow-beige | tin | decomposition |
| ***Polystichum acrostichoides*** CHRISTMAS FERN | | | |
| leaves | yellow-beige | copper | heat |
| | khaki-beige | iron | heat |
| ***Polytaenia nuttallii*** PRAIRIE PARSLEY | | | |
| flowers | yellow-beige | copper | heat |
| leaves and stems | yellow-beige | iron | decomposition |
| roots | yellow-beige | copper | heat |
| | khaki-beige | tin | heat |
| ***Populus deltoides*** EASTERN COTTONWOOD | | | |
| leaves | yellow-beige | iron | heat |
| | yellow-beige | iron | decomposition |
| roots | gray-beige | iron | decomposition |
| | beige | none | decomposition |
| stems | khaki-beige | copper | decomposition |
| | khaki-beige | iron | decomposition |
| ***Potentilla arguta*** TALL CINQUEFOIL | | | |
| flowers | yellow-beige | none | decomposition |
| leaves | yellow-beige | none | heat |
| rootstocks | tan-beige | alum | decomposition |

| DYE SOURCE | DYE RESULT | MORDANT | PROCESSING |
|---|---|---|---|
| ***Prosopis glandulosa*** MESQUITE | | | |
| roots | yellow-beige | alum | decomposition |
| | yellow-beige | copper | decomposition |
| | yellow-beige | none | decomposition |
| ***Psoralidium tenuiflorum*** SLIM-FLOWER SCURFPEA | | | |
| flowers | khaki-beige | copper | heat |
| leaves | light olive-beige | none | decomposition |
| roots | yellow-beige | none | heat |
| ***Ptilimnium nuttallii*** MOCK BISHOP'S-WEED | | | |
| roots | khaki-beige | copper | heat |
| ***Ratibida columnifera*** MEXICAN HAT | | | |
| ray florets | yellow-beige | iron | heat |
| roots | khaki-beige | copper | heat |
| | yellow-beige | tin | heat |
| ***Rhus copallinum*** WINGED SUMAC | | | |
| bark | yellow-beige | tin | decomposition |
| flowers | yellow-beige | none | heat |
| leaves | gray-beige | none | heat |
| | yellow-beige | none | decomposition |
| ***Rhus glabra*** SMOOTH SUMAC | | | |
| bark | yellow-beige | copper | decomposition |
| | gray-beige | iron | decomposition |
| | yellow-beige | tin | decomposition |
| ***Robinia hispida*** BRISTLY LOCUST | | | |
| flowers | yellow-beige | alum | heat |
| | yellow-beige | copper | heat |
| | yellow-beige | iron | decomposition |
| leaves | beige | none | decomposition |
| ***Robinia pseudoacacia*** BLACK LOCUST | | | |
| bark | yellow-beige | alum | heat |
| leaves | light olive-beige | alum | heat |
| | beige | alum | decomposition |
| petals | yellow-beige | copper | heat |
| ***Rudbeckia grandiflora*** ROUGH CONEFLOWER | | | |
| disk florets | yellow-beige | none | heat |
| leaves and stems | yellow-beige | tin | heat |
| | yellow-beige | tin | decomposition |
| ray florets | yellow-beige | copper | heat |
| | khaki-beige | iron | decomposition |
| | yellow-beige | none | decomposition |
| rhizomes | khaki-beige | copper | heat |
| | beige | copper | decomposition |
| | beige | iron | decomposition |
| | beige | none | decomposition |
| ***Rudbeckia hirta*** BLACK-EYED SUSAN | | | |
| ray florets | yellow-beige | iron | decomposition |
| roots | yellow-beige | copper | heat |
| ***Rumex altissimus*** PALE DOCK | | | |
| flowers | beige | iron | decomposition |

| DYE SOURCE | DYE RESULT | MORDANT | PROCESSING |
|---|---|---|---|
| ***Sabatia campestris*** PRAIRIE ROSE GENTIAN | | | |
| leaves and stems | yellow-beige | none | heat |
| petals | yellow-beige | copper | heat |
| | yellow-beige | iron | heat |
| ***Salix exigua*** SANDBAR WILLOW | | | |
| stems | yellow-beige | alum | heat |
| | tan-beige | alum | decomposition |
| | khaki-beige | copper | decomposition |
| | khaki-beige | iron | decomposition |
| | tan-beige | tin | decomposition |
| ***Salix nigra*** BLACK WILLOW | | | |
| bark | yellow-beige | copper | decomposition |
| branchlets | yellow-beige | none | heat |
| leaves | yellow-beige | copper | decomposition |
| ***Sambucus canadensis*** ELDERBERRY | | | |
| bark | yellow-beige | alum | heat |
| | khaki-beige | iron | heat |
| | tan-beige | none | heat |
| leaves | yellow-beige | alum | decomposition |
| | khaki-beige | copper | decomposition |
| | yellow-beige | iron | decomposition |
| | khaki-beige | none | heat |
| ***Sanguinaria canadensis*** BLOODROOT | | | |
| leaves and stems | yellow-beige | alum | decomposition |
| | khaki-beige | copper | heat |
| ***Sapindus drummondii*** SOAPBERRY | | | |
| bark | yellow-beige | copper | decomposition |
| | yellow-beige | iron | heat |
| | yellow-beige | iron | decomposition |
| | yellow-beige | tin | heat |
| leaves | khaki-beige | copper | decomposition |
| | light olive-beige | none | heat |
| stems | yellow-beige | iron | decomposition |
| | yellow-beige | none | heat |
| ***Sassafras albidum*** SASSAFRAS | | | |
| roots | yellow-beige | iron | decomposition |
| | gray-beige | none | heat |
| ***Sedum nuttallianum*** YELLOW STONECROP | | | |
| fruits, flowers, and stems | yellow-beige | copper | heat |
| ***Senecio riddellii*** SAND GROUNDSEL | | | |
| disk and ray florets | khaki-beige | alum | heat |
| | yellow-beige | alum | decomposition |
| | yellow-beige | copper | decomposition |
| | yellow-beige | none | decomposition |
| leaves | yellow-beige | none | heat |
| | yellow-beige | none | decomposition |
| roots | yellow-beige | alum | decomposition |
| | yellow-beige | copper | decomposition |
| | yellow-beige | iron | decomposition |
| | yellow-beige | none | decomposition |

| DYE SOURCE | DYE RESULT | MORDANT | PROCESSING |
|---|---|---|---|
| ***Silene stellata*** STARRY CAMPION | | | |
| stems (basal) and rootstocks | tan-beige | alum | decomposition |
| | yellow-beige | copper | heat |
| | beige | copper | decomposition |
| | yellow-beige | iron | heat |
| | gray-beige | iron | decomposition |
| | tan-beige | tin | decomposition |
| | tan-beige | none | decomposition |
| ***Silphium integrifolium*** PRAIRIE ROSINWEED | | | |
| leaves | yellow-beige | alum | decomposition |
| | khaki-beige | iron | decomposition |
| disk and ray florets | khaki-beige | copper | heat |
| ***Sisyrinchium angustifolium*** BLUE-EYED GRASS | | | |
| flowers | yellow-beige | alum | decomposition |
| | yellow-beige | copper | decomposition |
| | yellow-beige | none | decomposition |
| leaves | yellow-beige | alum | heat |
| ***Smilax bona-nox*** GREENBRIER | | | |
| stems and leaves | yellow-beige | copper | decomposition |
| | khaki-beige | iron | decomposition |
| | tan-beige | none | heat |
| ***Solanum dimidiatum*** WESTERN HORSE-NETTLE | | | |
| petals | yellow-beige | alum | decomposition |
| | yellow-beige | copper | heat |
| | yellow-beige | copper | decomposition |
| roots | khaki-beige | alum | decomposition |
| | khaki-beige | iron | decomposition |
| stamens | yellow-beige | copper | heat |
| ***Solanum elaeagnifolium*** SILVERLEAF NIGHTSHADE | | | |
| leaves | light olive-beige | alum | decomposition |
| | light gray-beige | none | decomposition |
| petals | khaki-beige | iron | heat |
| roots | beige | copper | heat |
| | yellow-beige | iron | heat |
| | yellow-beige | tin | heat |
| ***Solidago missouriensis*** PLAINS GOLDENROD | | | |
| leaves | light olive-beige | none | heat |
| rhizomes and roots | yellow-beige | alum | heat |
| | yellow-beige | copper | decomposition |
| | khaki-beige | iron | decomposition |
| stems | khaki-beige | none | heat |
| | yellow-beige | none | decomposition |
| ***Solidago rigida*** STIFF GOLDENROD | | | |
| heads | yellow-beige | none | heat |
| | yellow-beige | none | decomposition |
| roots | yellow-beige | alum | heat |
| | light olive-beige | copper | heat |
| | light gray-beige | none | decomposition |

| DYE SOURCE | DYE RESULT | MORDANT | PROCESSING |
|---|---|---|---|
| ***Sorghastrum nutans*** INDIANGRASS | | | |
| leaves | khaki-beige | iron | decomposition |
| rhizomes and roots | yellow-beige | copper | heat |
| spikelets | yellow-beige | copper | decomposition |
| stems | khaki-beige | copper | decomposition |
| ***Stillingia sylvatica*** QUEEN'S DELIGHT | | | |
| roots | yellow-beige | copper | decomposition |
| | gray-beige | iron | heat |
| stems | khaki-beige | iron | decomposition |
| ***Streptanthus hyacinthoides*** SMOOTH TWIST-FLOWER | | | |
| flowers | yellow-beige | copper | heat |
| | beige | iron | heat |
| ***Tephrosia virginiana*** GOAT'S RUE | | | |
| leaves | yellow-beige | copper | decomposition |
| | yellow-beige | iron | decomposition |
| | yellow-beige | none | heat |
| rhizomes | yellow-beige | alum | heat |
| | yellow-beige | copper | heat |
| | yellow-beige | copper | decomposition |
| | yellow-beige | iron | heat |
| | yellow-beige | iron | decomposition |
| | yellow-beige | none | heat |
| ***Teucrium canadense*** AMERICAN GERMANDER | | | |
| flowers | yellow-beige | iron | heat |
| roots | yellow-beige | alum | heat |
| | khaki-beige | copper | heat |
| ***Tradescantia occidentalis*** PRAIRIE SPIDERWORT | | | |
| stems | yellow-beige | copper | heat |
| ***Ulmus rubra*** SLIPPERY ELM | | | |
| bark | yellow-beige | alum | heat |
| | yellow-beige | copper | decomposition |
| | gray-beige | iron | decomposition |
| | tan-beige | none | heat |
| leaves | tan-beige | none | decomposition |
| ***Vaccinium arboreum*** FARKLEBERRY | | | |
| rootstocks | beige | copper | heat |
| stems | yellow-beige | alum | heat |
| | gray-beige | iron | heat |
| ***Vernonia baldwinii*** WESTERN IRONWEED | | | |
| flower buds | beige | none | heat |
| leaves | yellow-beige | iron | decomposition |
| ***Viburnum rufidulum*** RUSTY BLACK HAW | | | |
| leaves | khaki-beige | iron | decomposition |
| | yellow-beige | none | decomposition |
| pedicels | tan-beige | none | decomposition |
| roots | yellow-beige | alum | heat |
| | yellow-beige | alum | decomposition |
| | yellow-beige | copper | heat |

| DYE SOURCE | DYE RESULT | MORDANT | PROCESSING |
|---|---|---|---|
| | yellow-beige | copper | decomposition |
| | yellow-beige | none | heat |
| ***Vitis aestivalis*** PIGEON GRAPE | | | |
| bark | yellow-beige | copper | decomposition |
| fruits | gray-beige | alum | decomposition |
| | gray-beige | copper | decomposition |
| | gray-beige | iron | decomposition |
| | gray-beige | none | decomposition |
| leaves | yellow-beige | copper | heat |
| | yellow-beige | copper | decomposition |
| | yellow-beige | iron | decomposition |

Our search for natural dyes resulted in various cream colors, shown here with Munsell notations. Following these color chips is a list of the plants that were used to obtain these colors, including mordant and processing conditions.

5Y 9/2 7.5Y 9/2 10Y 9/2

**CREAM**

2.5GY 9/2

**PALE GREEN·CREAM**

5Y 9/4

**PALE YELLOW·CREAM**

5YR 9/1 10YR 9/1

**PINK·CREAM**

| DYE SOURCE | DYE RESULT | MORDANT | PROCESSING |
|---|---|---|---|
| ***Acer negundo*** BOXELDER | | | |
| bark | cream | alum | decomposition |
| | pale yellow-cream | copper | decomposition |
| | pale yellow-cream | iron | decomposition |
| | cream | none | decomposition |
| leaves | pale yellow-cream | alum | heat |
| | pale yellow-cream | alum | decomposition |
| | pale yellow-cream | copper | decomposition |
| | pale yellow-cream | none | heat |
| | pale yellow-cream | none | decomposition |
| ***Acer saccharinum*** SILVER MAPLE | | | |
| leaves | pale yellow-cream | none | heat |
| ***Achillea millefolium*** YARROW | | | |
| flowers | pale yellow-cream | copper | decomposition |
| ***Ageratina altissima*** WHITE SNAKEROOT | | | |
| roots | cream | none | heat |
| stems | pale yellow-cream | copper | decomposition |
| | pale yellow-cream | iron | decomposition |
| | pale yellow-cream | none | heat |
| ***Ambrosia psilostachya*** WESTERN RAGWEED | | | |
| leaves and heads | pale yellow-cream | tin | decomposition |
| | cream | none | decomposition |
| rhizomes and roots | pale yellow-cream | alum | decomposition |
| | cream | none | decomposition |
| ***Ambrosia trifida*** GIANT RAGWEED | | | |
| leaves | cream | alum | decomposition |
| | pale yellow-cream | copper | decomposition |
| | pale yellow-cream | iron | decomposition |
| | pale yellow-cream | none | decomposition |
| roots | pale yellow-cream | alum | heat |
| | cream | none | heat |
| | cream | none | decomposition |
| stems | pale green-cream | none | heat |
| ***Amorpha canescens*** LEADPLANT | | | |
| bark | pale yellow-cream | alum | decomposition |
| | pale yellow-cream | tin | decomposition |
| | pale yellow-cream | none | decomposition |
| leaves | pale yellow-cream | none | heat |
| spikes | cream | none | decomposition |
| ***Apocynum cannabinum*** INDIAN HEMP | | | |
| roots | cream | none | decomposition |
| stems | pale yellow-cream | tin | decomposition |
| ***Argemone polyanthemos*** PRICKLY POPPY | | | |
| leaves | pale yellow-cream | alum | decomposition |
| | pale yellow-cream | none | heat |
| | pale yellow-cream | none | decomposition |
| pistils | pale yellow-cream | alum | heat |
| | pale yellow-cream | copper | decomposition |
| | cream | none | heat |
| | cream | none | decomposition |

| DYE SOURCE | DYE RESULT | MORDANT | PROCESSING |
|---|---|---|---|
| roots | cream | alum | decomposition |
| | cream | copper | decomposition |
| stems | pale yellow-cream | alum | heat |
| | pale yellow-cream | alum | decomposition |
| | pale yellow-cream | copper | decomposition |
| | pale yellow-cream | iron | decomposition |
| | pale yellow-cream | none | heat |
| | pale yellow-cream | none | decomposition |
| ***Asclepias arenaria*** SAND MILKWEED | | | |
| flowers | cream | none | heat |
| fruits | cream | alum | heat |
| | cream | alum | decomposition |
| | cream | copper | heat |
| | cream | copper | decomposition |
| | pale yellow-cream | iron | heat |
| | cream | iron | decomposition |
| | cream | none | heat |
| | cream | none | decomposition |
| leaves | pale yellow-cream | tin | decomposition |
| roots | cream | alum | heat |
| | cream | copper | heat |
| | cream | iron | heat |
| | pale yellow-cream | tin | heat |
| | cream | none | heat |
| stems | cream | alum | heat |
| | cream | alum | decomposition |
| | cream | copper | heat |
| | cream | copper | decomposition |
| | cream | iron | heat |
| | cream | iron | decomposition |
| | cream | tin | decomposition |
| | cream | none | heat |
| | cream | none | decomposition |
| ***Asclepias tuberosa*** BUTTERFLY MILKWEED | | | |
| flowers | cream | none | heat |
| leaves | pale green-cream | alum | decomposition |
| | cream | none | decomposition |
| ***Asclepias viridis*** ANTELOPE-HORN MILKWEED | | | |
| flowers | cream | alum | heat |
| | pale yellow-cream | copper | heat |
| | pale yellow-cream | iron | heat |
| | cream | none | heat |
| | cream | none | decomposition |
| roots | pale yellow-cream | alum | heat |
| | cream | alum | decomposition |
| | cream | copper | decomposition |
| | pale yellow-cream | iron | decomposition |
| | cream | tin | decomposition |
| | cream | none | heat |
| | cream | none | decomposition |

| DYE SOURCE | DYE RESULT | MORDANT | PROCESSING |
|---|---|---|---|
| ***Asimina triloba*** PAWPAW | | | |
| leaves | cream | tin | decomposition |
| | cream | none | heat |
| rootstocks immature | pale yellow-cream | tin | heat |
| | cream | none | heat |
| ***Baccharis salicina*** WILLOW BACCHARIS | | | |
| leaves | pale yellow-cream | copper | decomposition |
| | pale yellow-cream | none | decomposition |
| roots | cream | alum | decomposition |
| | pale yellow-cream | tin | heat |
| | pale yellow-cream | tin | decomposition |
| | cream | none | heat |
| ***Baptisia australis*** BLUE WILD-INDIGO | | | |
| petals | pale yellow-cream | alum | decomposition |
| | cream | copper | decomposition |
| | cream | none | decomposition |
| roots | pale yellow-cream | tin | decomposition |
| stems (basal) | cream | none | heat |
| ***Baptisia sphaerocarpa*** GOLDEN WILD-INDIGO | | | |
| flowers | cream | none | heat |
| fruits and bracts | pale yellow-cream | tin | decomposition |
| leaves | pale yellow-cream | alum | heat |
| | pale yellow-cream | none | heat |
| roots | pale yellow-cream | alum | decomposition |
| | pale yellow-cream | tin | decomposition |
| stems | pale yellow-cream | copper | heat |
| | pale yellow-cream | tin | decomposition |
| | pale yellow-cream | none | heat |
| ***Bidens aristosa*** BEARDED BEGGAR-TICKS | | | |
| stems | pale yellow-cream | none | heat |
| ***Bidens bipinnata*** SPANISH NEEDLES | | | |
| leaves | cream | alum | decomposition |
| | cream | iron | decomposition |
| | cream | tin | decomposition |
| | cream | none | decomposition |
| roots | cream | alum | decomposition |
| | cream | iron | decomposition |
| | cream | tin | decomposition |
| | pale yellow-cream | none | heat |
| | cream | none | decomposition |
| stems | cream | alum | decomposition |
| | cream | copper | decomposition |
| | cream | iron | decomposition |
| | cream | tin | decomposition |
| | cream | none | decomposition |
| ***Brickellia eupatorioides*** FALSE BONESET | | | |
| roots | pale green-cream | tin | decomposition |
| stems | pale yellow-cream | alum | decomposition |

| DYE SOURCE | DYE RESULT | MORDANT | PROCESSING |
|---|---|---|---|
| | cream | none | heat |
| | pale yellow-cream | none | decomposition |
| ***Bumelia lanuginosa*** CHITTAMWOOD | | | |
| branchlets | pale yellow-cream | alum | heat |
| | cream | alum | decomposition |
| | pale yellow-cream | tin | decomposition |
| | pale yellow-cream | none | heat |
| | cream | none | decomposition |
| fruits | cream | none | decomposition |
| leaves | cream | none | decomposition |
| ***Callicarpa americana*** AMERICAN BEAUTYBERRY | | | |
| fruits | cream | none | heat |
| ***Calylophus hartwegii*** HARTWEG'S SUNDROP | | | |
| leaves | pale yellow-cream | none | heat |
| | pale yellow-cream | none | decomposition |
| petals | pale yellow-cream | alum | heat |
| roots | cream | alum | heat |
| | cream | alum | decomposition |
| | cream | tin | decomposition |
| ***Calylophus serrulatus*** HALFSHRUB SUNDROP | | | |
| petals | pale yellow-cream | alum | heat |
| stems | pale yellow-cream | alum | decomposition |
| | cream | none | decomposition |
| ***Campsis radicans*** TRUMPET CREEPER | | | |
| flowers | cream | none | heat |
| fruits | pale yellow-cream | alum | heat |
| | cream | none | heat |
| leaves | pale yellow-cream | tin | decomposition |
| sepals | cream | none | heat |
| ***Carya cordiformis*** BITTERNUT HICKORY | | | |
| fruits | pale yellow-cream | none | heat |
| | pale yellow-cream | none | decomposition |
| leaves | pale yellow-cream | none | heat |
| ***Castanea ozarkensis*** OZARK CHINKAPIN | | | |
| leaves | pale yellow-cream | none | decomposition |
| ***Castilleja indivisa*** INDIAN PAINTBRUSH | | | |
| bract tips | pale yellow-cream | copper | heat |
| ***Castilleja purpurea* var. *citrina*** YELLOW PAINTBRUSH | | | |
| leaves | pale yellow-cream | none | heat |
| petals and bracts | pale yellow-cream | none | heat |
| ***Ceanothus americanus*** NEW JERSEY TEA | | | |
| flowers | cream | alum | decomposition |
| | cream | none | heat |
| | cream | none | decomposition |
| leaves | cream | none | heat |
| rhizomes and roots | pale yellow-cream | tin | decomposition |
| stems | pale yellow-cream | alum | decomposition |
| | pale yellow-cream | tin | decomposition |

| DYE SOURCE | DYE RESULT | MORDANT | PROCESSING |
|---|---|---|---|
| ***Centaurea americana*** BASKET FLOWER | | | |
| disk florets | cream | alum | heat |
| | cream | copper | heat |
| | cream | iron | heat |
| | cream | none | heat |
| ***Cephalanthus occidentalis*** BUTTONBUSH | | | |
| flowers | pale yellow-cream | alum | heat |
| | cream | alum | decomposition |
| | cream | copper | decomposition |
| | pale yellow-cream | none | heat |
| | cream | none | decomposition |
| leaves | cream | alum | decomposition |
| | cream | copper | decomposition |
| | cream | none | decomposition |
| | pale yellow-cream | alum | heat |
| | pale yellow-cream | none | heat |
| ***Cercis canadensis*** REDBUD | | | |
| flowers | pale yellow-cream | none | heat |
| ***Chrysopsis pilosa*** SOFT GOLDEN-ASTER | | | |
| leaves | cream | none | decomposition |
| petals | cream | none | decomposition |
| ray florets | pale yellow-cream | none | heat |
| roots | cream | none | heat |
| ***Cicuta maculata*** WATER-HEMLOCK | | | |
| flowers | pale yellow-cream | copper | decomposition |
| | cream | none | decomposition |
| leaves | pale yellow-cream | copper | decomposition |
| | pale yellow-cream | none | decomposition |
| roots | cream | alum | heat |
| | cream | tin | heat |
| | cream | tin | decomposition |
| | cream | none | heat |
| stems | pale yellow-cream | alum | heat |
| | cream | alum | decomposition |
| | cream | copper | decomposition |
| | cream | iron | decomposition |
| | cream | tin | decomposition |
| | pale yellow-cream | none | heat |
| | cream | none | decomposition |
| ***Cirsium undulatum*** WAVY-LEAF THISTLE | | | |
| leaves | pale yellow-cream | tin | decomposition |
| pappus | pale yellow-cream | alum | heat |
| | cream | alum | decomposition |
| | pale yellow-cream | copper | decomposition |
| | pale yellow-cream | none | heat |
| | pale yellow-cream | none | decomposition |
| ***Comandra umbellata*** BASTARD TOAD-FLAX | | | |
| roots | cream | alum | heat |
| | cream | copper | heat |

| DYE SOURCE | DYE RESULT | MORDANT | PROCESSING |
|---|---|---|---|
| | cream | iron | heat |
| | cream | tin | heat |
| | cream | none | heat |
| ***Cornus drummondii*** ROUGH-LEAF DOGWOOD | | | |
| fruits | cream | none | heat |
| | cream | none | decomposition |
| leaves | pale yellow-cream | none | heat |
| | pale yellow-cream | none | decomposition |
| pedicels | pale yellow-cream | alum | decomposition |
| | cream | none | heat |
| | pale yellow-cream | none | decomposition |
| roots | cream | none | decomposition |
| ***Cotinus obovatus*** SMOKE-TREE | | | |
| bark | pale yellow-cream | alum | decomposition |
| | pale yellow-cream | none | decomposition |
| leaves | pale yellow-cream | none | decomposition |
| ***Cucurbita foetidissima*** BUFFALO GOURD | | | |
| fruits | cream | alum | heat |
| | pale yellow-cream | alum | decomposition |
| | pale yellow-cream | copper | heat |
| | pale yellow-cream | copper | decomposition |
| | pale yellow-cream | iron | heat |
| | pale yellow-cream | iron | decomposition |
| | pale yellow-cream | tin | decomposition |
| | cream | none | heat |
| | pale yellow-cream | none | decomposition |
| leaves | pale yellow-cream | none | heat |
| | pale yellow-cream | none | decomposition |
| stems | pale yellow-cream | alum | heat |
| | cream | none | heat |
| ***Cycloloma atriplicifolium*** TUMBLE RINGWING | | | |
| leaves and inflorescences | cream | alum | decomposition |
| | pale yellow-cream | none | heat |
| | cream | none | decomposition |
| roots | cream | alum | heat |
| | cream | copper | heat |
| | cream | iron | heat |
| | cream | tin | heat |
| | cream | none | heat |
| stems | cream | alum | decomposition |
| | pale yellow-cream | copper | decomposition |
| | pale yellow-cream | iron | decomposition |
| | pale yellow-cream | none | decomposition |
| ***Cyperus squarrosus*** BEARDED FLATSEDGE | | | |
| spikes | pale yellow-cream | alum | decomposition |
| ***Dalea candida*** WHITE PRAIRIE-CLOVER | | | |
| flowers | pale yellow-cream | alum | heat |
| | cream | none | heat |
| | pale yellow-cream | none | decomposition |

| DYE SOURCE | DYE RESULT | MORDANT | PROCESSING |
|---|---|---|---|
| ***Dalea lanata*** WOOLLY DALEA | | | |
| roots | pale yellow-cream | alum | decomposition |
| | pale yellow-cream | copper | decomposition |
| | pale yellow-cream | tin | decomposition |
| | pale yellow-cream | none | heat |
| | pale yellow-cream | none | decomposition |
| stems | cream | alum | decomposition |
| | cream | copper | decomposition |
| | pale yellow-cream | iron | decomposition |
| | cream | none | decomposition |
| ***Dalea purpurea*** PURPLE PRAIRIE-CLOVER | | | |
| flowers | cream | alum | heat |
| | cream | none | heat |
| leaves and stems | cream | none | heat |
| ***Dalea villosa*** SILKY PRAIRIE-CLOVER | | | |
| flowers | cream | none | heat |
| | cream | none | decomposition |
| leaves | cream | none | decomposition |
| roots | cream | tin | decomposition |
| | cream | none | decomposition |
| stems | cream | none | heat |
| | cream | none | decomposition |
| ***Desmanthus illinoensis*** ILLINOIS BUNDLEFLOWER | | | |
| fruits | pale yellow-cream | alum | decomposition |
| | cream | none | decomposition |
| leaves | cream | none | decomposition |
| stems | cream | none | heat |
| | cream | none | decomposition |
| ***Desmodium glutinosum*** STICKY TICKCLOVER | | | |
| leaves | cream | alum | decomposition |
| | pale yellow-cream | tin | decomposition |
| | cream | none | heat |
| | cream | none | decomposition |
| stems | cream | alum | decomposition |
| | pale yellow-cream | tin | decomposition |
| ***Diospyros virginiana*** PERSIMMON | | | |
| bark | pale yellow-cream | alum | decomposition |
| | pale yellow-cream | iron | decomposition |
| | pale yellow-cream | tin | decomposition |
| | pale yellow-cream | none | decomposition |
| branchlets | pale yellow-cream | none | decomposition |
| leaves | pale yellow-cream | alum | decomposition |
| fruits | pale yellow-cream | copper | decomposition |
| | pale yellow-cream | iron | decomposition |
| | cream | none | heat |
| | pale yellow-cream | none | decomposition |
| ***Dracopis amplexicaulis*** CLASPING-LEAVED CONEFLOWER | | | |
| disk florets | pink-cream | alum | decomposition |
| | pink-cream | none | decomposition |

| DYE SOURCE | DYE RESULT | MORDANT | PROCESSING |
|---|---|---|---|
| leaves | pale yellow-cream | tin | decomposition |
| ray florets | cream | alum | decomposition |
| | pale yellow-cream | copper | decomposition |
| | cream | none | decomposition |
| roots | cream | alum | decomposition |
| | cream | iron | decomposition |
| | cream | tin | decomposition |
| stems | pale yellow-cream | alum | heat |
| | cream | tin | decomposition |
| | cream | none | heat |
| ***Echinacea angustifolia*** PURPLE PRAIRIE CONEFLOWER | | | |
| ray florets | pale yellow-cream | none | heat |
| roots | pale yellow-cream | alum | heat |
| | cream | none | heat |
| ***Echinocereus reichenbachii*** LACE HEDGEHOG-CACTUS | | | |
| flowers | cream | alum | decomposition |
| | pale yellow-cream | none | heat |
| ***Elephantopus carolinianus*** LEAFY ELEPHANT'S FOOT | | | |
| leaves | cream | none | heat |
| roots | cream | none | heat |
| stems | cream | none | heat |
| ***Euphorbia maculata*** SPOTTED SPURGE | | | |
| leaves and stems | cream | none | heat |
| ***Euphorbia marginata*** SNOW-ON-THE-MOUNTAIN | | | |
| bracts | pale yellow-cream | alum | decomposition |
| | cream | none | decomposition |
| leaves | pale yellow-cream | none | decomposition |
| roots | cream | alum | decomposition |
| | cream | none | heat |
| stems | pale yellow-cream | alum | decomposition |
| | pale yellow-cream | tin | decomposition |
| | cream | none | heat |
| ***Eustoma exaltatum*** PRAIRIE GENTIAN | | | |
| petals | pale yellow-cream | none | decomposition |
| stems | pale yellow-cream | alum | decomposition |
| | pale yellow-cream | copper | decomposition |
| | pale yellow-cream | tin | decomposition |
| | pale yellow-cream | none | decomposition |
| ***Fraxinus americana*** WHITE ASH | | | |
| fruits | cream | none | heat |
| ***Fraxinus pennsylvanica*** GREEN ASH | | | |
| bark | pale yellow-cream | alum | decomposition |
| | pale yellow-cream | copper | decomposition |
| | cream | none | heat |
| | pale yellow-cream | none | decomposition |
| fruits | cream | alum | heat |
| | cream | alum | decomposition |
| | cream | copper | heat |
| | cream | iron | heat |

| DYE SOURCE | DYE RESULT | MORDANT | PROCESSING |
|---|---|---|---|
| ***Fraxinus pennsylvanica,*** fruits, continued | | | |
| | cream | tin | heat |
| | cream | none | heat |
| | cream | none | decomposition |
| ***Fraxinus quadrangulata*** BLUE ASH | | | |
| bark | pale yellow-cream | none | decomposition |
| leaves | pale yellow-cream | none | heat |
| ***Froelichia floridana*** FIELD SNAKE-COTTON | | | |
| flowers | cream | alum | decomposition |
| | cream | copper | decomposition |
| | pale yellow-cream | tin | decomposition |
| | cream | none | decomposition |
| leaves | pale yellow-cream | alum | decomposition |
| | cream | none | heat |
| roots | cream | alum | heat |
| | cream | alum | decomposition |
| | cream | copper | heat |
| | cream | copper | decomposition |
| | cream | iron | heat |
| | cream | iron | decomposition |
| | cream | tin | heat |
| | cream | tin | decomposition |
| | cream | none | heat |
| | cream | none | decomposition |
| stems | cream | alum | decomposition |
| | cream | copper | decomposition |
| | cream | iron | decomposition |
| | cream | tin | decomposition |
| | pale yellow-cream | none | heat |
| | cream | none | decomposition |
| ***Gaillardia aestivalis*** PRAIRIE GAILLARDIA | | | |
| ray florets | cream | none | heat |
| roots | pale yellow-cream | alum | heat |
| | cream | none | heat |
| ***Gaillardia pulchella*** INDIAN BLANKET | | | |
| disk florets | pale yellow-cream | alum | decomposition |
| | pale yellow-cream | copper | decomposition |
| | pale yellow-cream | none | heat |
| | pale yellow-cream | none | decomposition |
| roots | pale yellow-cream | alum | heat |
| | cream | alum | decomposition |
| | pale yellow-cream | copper | heat |
| | pale yellow-cream | iron | heat |
| | pale yellow-cream | tin | heat |
| | cream | tin | decomposition |
| | pale yellow-cream | none | heat |
| ***Glandularia canadensis*** ROSE VERVAIN | | | |
| flowers | cream | none | heat |

| DYE SOURCE | DYE RESULT | MORDANT | PROCESSING |
|---|---|---|---|
| ***Gleditsia triacanthos*** HONEY LOCUST | | | |
| fruits | pale yellow-cream | alum | heat |
| | pale yellow-cream | alum | decomposition |
| | pale yellow-cream | copper | decomposition |
| | pale yellow-cream | iron | decomposition |
| | pale yellow-cream | none | heat |
| | pale yellow-cream | none | decomposition |
| ***Gymnocladus dioicus*** KENTUCKY COFFEE-TREE | | | |
| bark | cream | none | decomposition |
| fruits | pale yellow-cream | alum | heat |
| | pale yellow-cream | iron | heat |
| | cream | none | heat |
| | pale yellow-cream | none | decomposition |
| leaflets | pale yellow-cream | copper | decomposition |
| | pale yellow-cream | iron | decomposition |
| petioles | cream | alum | decomposition |
| | pale yellow-cream | iron | decomposition |
| | cream | none | decomposition |
| seeds | pale yellow-cream | alum | heat |
| | pale yellow-cream | iron | decomposition |
| | cream | none | heat |
| stems | pale yellow-cream | alum | heat |
| | pale yellow-cream | alum | decomposition |
| | pale yellow-cream | tin | decomposition |
| | pale yellow-cream | none | decomposition |
| ***Helenium amarum*** BITTER SNEEZEWEED | | | |
| flowers | pale yellow-cream | none | decomposition |
| roots | pale yellow-cream | alum | decomposition |
| | pale yellow-cream | iron | decomposition |
| | pale yellow-cream | none | decomposition |
| stems | pale yellow-cream | alum | decomposition |
| | pale yellow-cream | tin | decomposition |
| | cream | none | heat |
| | pale yellow-cream | none | decomposition |
| ***Helianthus annuus*** ANNUAL SUNFLOWER | | | |
| disk florets | cream | iron | heat |
| | cream | none | heat |
| ray florets | cream | none | heat |
| ***Helianthus maximiliani*** MAXIMILIAN'S SUNFLOWER | | | |
| ray florets | pale yellow-cream | none | decomposition |
| ***Helianthus tuberosus*** JERUSALEM ARTICHOKE | | | |
| disk and ray florets | pale yellow-cream | copper | decomposition |
| roots | cream | alum | decomposition |
| | pale yellow-cream | tin | decomposition |
| | cream | none | heat |
| stems | cream | none | heat |
| ***Hibiscus laevis*** SCARLET ROSE MALLOW | | | |
| leaves | cream | none | heat |
| petals | pale yellow-cream | none | heat |

| DYE SOURCE | DYE RESULT | MORDANT | PROCESSING |
|---|---|---|---|
| ***Hibiscus laevis,*** petals, continued | | | |
| | cream | none | decomposition |
| sepals | cream | alum | heat |
| | cream | alum | decomposition |
| | pale yellow-cream | copper | heat |
| | cream | tin | decomposition |
| | cream | none | heat |
| | cream | none | decomposition |
| stems | cream | alum | decomposition |
| | cream | copper | decomposition |
| | cream | iron | decomposition |
| | cream | tin | decomposition |
| | cream | none | heat |
| | cream | none | decomposition |
| ***Hydrangea arborescens*** WILD HYDRANGEA | | | |
| branches | cream | alum | heat |
| | cream | alum | decomposition |
| | cream | copper | decomposition |
| | cream | iron | decomposition |
| | pale yellow-cream | tin | heat |
| | pale yellow-cream | tin | decomposition |
| | cream | none | heat |
| | cream | none | decomposition |
| flowers | pale yellow-cream | none | heat |
| | cream | none | decomposition |
| leaves | pale yellow-cream | alum | heat |
| | cream | alum | decomposition |
| | pale yellow-cream | copper | decomposition |
| | pale yellow-cream | none | heat |
| ***Ipomoea leptophylla*** BUSH MORNING-GLORY | | | |
| rootstocks | pale yellow-cream | alum | heat |
| | cream | alum | decomposition |
| | cream | copper | decomposition |
| | pale yellow-cream | none | heat |
| | cream | none | decomposition |
| stems | cream | alum | decomposition |
| | cream | copper | decomposition |
| | cream | iron | decomposition |
| | cream | none | heat |
| | cream | none | decomposition |
| ***Ipomopsis rubra*** STANDING CYPRESS | | | |
| petals | pale yellow-cream | none | decomposition |
| roots | pale yellow-cream | alum | heat |
| | cream | alum | decomposition |
| | cream | copper | decomposition |
| | pale yellow-cream | iron | heat |
| | cream | iron | decomposition |
| | cream | none | heat |
| | cream | none | decomposition |
| stems | cream | alum | decomposition |

| DYE SOURCE | DYE RESULT | MORDANT | PROCESSING |
|---|---|---|---|
| | cream | copper | decomposition |
| | cream | iron | decomposition |
| | cream | none | heat |
| | cream | none | decomposition |
| ***Juniperus virginiana*** EASTERN RED CEDAR | | | |
| bark | cream | alum | decomposition |
| | pale yellow-cream | iron | decomposition |
| | pale yellow-cream | tin | decomposition |
| | cream | none | decomposition |
| leaves and stems | pale yellow-cream | none | decomposition |
| ***Lesquerella gracilis*** SPREADING BLADDERPOD | | | |
| leaves and stems | pale yellow-cream | alum | heat |
| | pale yellow-cream | none | heat |
| ***Liatris punctata*** DOTTED GAYFEATHER | | | |
| disk florets | cream | alum | heat |
| | pale yellow-cream | copper | decomposition |
| | pale yellow-cream | iron | decomposition |
| | cream | none | heat |
| leaves and stems | cream | none | heat |
| phyllaries | pale yellow-cream | none | heat |
| roots | pale yellow-cream | alum | heat |
| | pale yellow-cream | alum | decomposition |
| | pale yellow-cream | copper | heat |
| | cream | none | heat |
| | cream | none | decomposition |
| ***Liatris squarrosa*** SCALY GAYFEATHER | | | |
| flowers | cream | none | heat |
| roots | pale yellow-cream | alum | heat |
| | cream | none | heat |
| ***Linum sulcatum*** GROOVED FLAX | | | |
| roots | cream | alum | heat |
| | pale yellow-cream | copper | heat |
| | cream | none | heat |
| ***Liquidambar styraciflua*** SWEET GUM | | | |
| bark | cream | none | decomposition |
| roots | pale yellow-cream | none | heat |
| ***Maclura pomifera*** OSAGE ORANGE | | | |
| bark | cream | none | decomposition |
| ***Mentzelia nuda*** BRACTLESS BLAZING-STAR | | | |
| flowers | cream | none | heat |
| leaves | cream | alum | heat |
| | cream | alum | decomposition |
| | pale yellow-cream | tin | decomposition |
| | cream | none | heat |
| | cream | none | decomposition |
| roots | cream | alum | decomposition |
| | cream | copper | decomposition |
| | pale yellow-cream | iron | decomposition |
| | cream | tin | decomposition |

| DYE SOURCE | DYE RESULT | MORDANT | PROCESSING |
|---|---|---|---|
| ***Mentzelia nuda,*** roots, continued | | | |
| | cream | none | decomposition |
| stems | cream | alum | decomposition |
| | cream | iron | decomposition |
| | cream | tin | decomposition |
| | cream | none | decomposition |
| ***Monarda fistulosa*** WILD BERGAMOT | | | |
| flowers | cream | none | decomposition |
| ***Morus rubra*** RED MULBERRY | | | |
| bark | pale yellow-cream | tin | decomposition |
| | cream | none | heat |
| bark green | pale yellow-cream | tin | decomposition |
| | pale yellow-cream | none | decomposition |
| branches | cream | alum | heat |
| | pale yellow-cream | alum | decomposition |
| | pale green-cream | copper | heat |
| | cream | iron | heat |
| | pale yellow-cream | iron | decomposition |
| | pale yellow-cream | tin | decomposition |
| | cream | none | heat |
| | cream | none | decomposition |
| branchlets | pale yellow-cream | alum | decomposition |
| | pale yellow-cream | none | heat |
| | pale yellow-cream | none | decomposition |
| leaves | cream | none | heat |
| | pale yellow-cream | none | decomposition |
| ***Nuttallanthus canadensis*** TOAD-FLAX | | | |
| leaves and stems | cream | alum | decomposition |
| | pale yellow-cream | tin | decomposition |
| | cream | none | decomposition |
| ***Nyssa sylvatica*** BLACK GUM | | | |
| bark | pale yellow-cream | alum | heat |
| branchlets | cream | alum | heat |
| | pale yellow-cream | alum | decomposition |
| | pale yellow-cream | tin | heat |
| | cream | none | heat |
| leaves | cream | none | heat |
| ***Oenothera heterophylla*** SAND EVENING-PRIMROSE | | | |
| flowers | pale yellow-cream | none | decomposition |
| roots | pale yellow-cream | tin | heat |
| ***Oenothera macrocarpa*** BIGFRUIT EVENING-PRIMROSE | | | |
| bracts and stems | cream | none | decomposition |
| petals | cream | tin | decomposition |
| | cream | none | decomposition |
| roots | pale yellow-cream | tin | heat |
| fruits mature | pale yellow-cream | alum | heat |
| | cream | none | heat |
| fruits immature | pale yellow-cream | alum | heat |

| DYE SOURCE | DYE RESULT | MORDANT | PROCESSING |
|---|---|---|---|
| ***Opuntia macrorhiza*** PRICKLY-PEAR | | | |
| fruits | pale yellow-cream | alum | heat |
| | pale yellow-cream | none | heat |
| ***Packera obovata*** ROUNDLEAF GROUNDSEL | | | |
| flowers | pale yellow-cream | copper | decomposition |
| | pale yellow-cream | iron | decomposition |
| | cream | none | decomposition |
| ***Parthenocissus quinquefolia*** VIRGINIA CREEPER | | | |
| fruits | pale yellow-cream | none | heat |
| leaves | pale yellow-cream | none | heat |
| pedicels | pale yellow-cream | alum | decomposition |
| | pale yellow-cream | tin | decomposition |
| | cream | none | heat |
| | pale yellow-cream | none | decomposition |
| ***Pediomelum cuspidatum*** TALL-BREAD SCURFPEA | | | |
| fruits | pale yellow-cream | alum | heat |
| | pale yellow-cream | alum | decomposition |
| | cream | none | heat |
| | cream | none | decomposition |
| roots | cream | alum | heat |
| | pale yellow-cream | copper | heat |
| | pale yellow-cream | iron | heat |
| | cream | none | heat |
| stems | pale yellow-cream | alum | heat |
| | pale yellow-cream | copper | heat |
| | pale yellow-cream | copper | decomposition |
| | pale yellow-cream | iron | decomposition |
| | cream | none | heat |
| ***Phlox pilosa*** PRAIRIE PHLOX | | | |
| leaves | pale yellow-cream | alum | decomposition |
| | pale yellow-cream | copper | decomposition |
| | pale yellow-cream | none | decomposition |
| petals | pale yellow-cream | none | heat |
| stems | pale yellow-cream | alum | heat |
| | cream | tin | decomposition |
| | pale yellow-cream | none | heat |
| | cream | none | decomposition |
| ***Phyla lanceolata*** LANCELEAF FROG-FRUIT | | | |
| flowers | pale yellow-cream | none | heat |
| stems | cream | alum | decomposition |
| | cream | tin | decomposition |
| | cream | none | decomposition |
| ***Physalis angulata*** CUTLEAF GROUND-CHERRY | | | |
| fruits | cream | alum | heat |
| | cream | alum | decomposition |
| | cream | copper | heat |
| | cream | copper | decomposition |
| | cream | iron | heat |
| | cream | tin | decomposition |

| DYE SOURCE | DYE RESULT | MORDANT | PROCESSING |
|---|---|---|---|
| ***Physalis angulata,*** fruits, continued | | | |
| | cream | none | heat |
| | cream | none | decomposition |
| leaves | cream | none | heat |
| roots | cream | alum | heat |
| | cream | alum | decomposition |
| | cream | copper | heat |
| | cream | copper | decomposition |
| | cream | iron | heat |
| | cream | iron | decomposition |
| | cream | tin | heat |
| | cream | tin | decomposition |
| | cream | none | heat |
| | cream | none | decomposition |
| stems | cream | alum | heat |
| | cream | alum | decomposition |
| | cream | copper | heat |
| | cream | copper | decomposition |
| | cream | iron | heat |
| | cream | iron | decomposition |
| | cream | none | heat |
| | cream | none | decomposition |
| ***Physalis virginiana*** VIRGINIA GROUND-CHERRY | | | |
| stems | cream | alum | heat |
| | pale yellow-cream | copper | heat |
| | cream | iron | heat |
| | cream | none | heat |
| ***Phytolacca americana*** POKEWEED | | | |
| roots | cream | alum | heat |
| | pale yellow-cream | alum | decomposition |
| | cream | copper | heat |
| | pale yellow-cream | copper | decomposition |
| | cream | iron | heat |
| | pale yellow-cream | tin | heat |
| | cream | none | heat |
| | pale yellow-cream | none | decomposition |
| stems | cream | alum | heat |
| | cream | copper | heat |
| | pale yellow-cream | copper | decomposition |
| | cream | iron | heat |
| | cream | none | heat |
| | cream | none | decomposition |
| ***Podophyllum peltatum*** MAY-APPLE | | | |
| leaves | pale yellow-cream | alum | heat |
| | pale yellow-cream | none | heat |
| roots and rhizomes | cream | alum | decomposition |
| ***Polygala alba*** WHITE MILKWORT | | | |
| flowers | cream | alum | heat |
| | pale yellow-cream | copper | heat |
| | pale yellow-cream | iron | heat |

| DYE SOURCE | DYE RESULT | MORDANT | PROCESSING |
|---|---|---|---|
| | cream | none | heat |
| | cream | none | decomposition |
| leaves and stems | pale yellow-cream | alum | heat |
| | pale yellow-cream | copper | heat |
| | pale yellow-cream | none | heat |
| ***Polygonum pensylvanicum*** PENNSYLVANIA SMARTWEED | | | |
| flowers | pale yellow-cream | none | decomposition |
| leaves | cream | alum | decomposition |
| stems | cream | alum | decomposition |
| | cream | none | heat |
| ***Polygonum punctatum*** DOTTED SMARTWEED | | | |
| flowers and flower buds | pale yellow-cream | alum | decomposition |
| | pale yellow-cream | iron | decomposition |
| | cream | none | decomposition |
| stems | cream | none | heat |
| | cream | none | decomposition |
| ***Polystichum acrostichoides*** CHRISTMAS FERN | | | |
| leaves | cream | alum | heat |
| | cream | alum | decomposition |
| | cream | copper | decomposition |
| | cream | iron | decomposition |
| | cream | none | heat |
| | cream | none | decomposition |
| ***Polytaenia nuttallii*** PRAIRIE PARSLEY | | | |
| roots | pale yellow-cream | alum | heat |
| | pale yellow-cream | alum | decomposition |
| | pale yellow-cream | none | heat |
| | cream | none | decomposition |
| ***Populus deltoides*** EASTERN COTTONWOOD | | | |
| leaves | pale yellow-cream | copper | decomposition |
| | cream | none | decomposition |
| roots | cream | alum | decomposition |
| | cream | tin | decomposition |
| | cream | none | heat |
| stems | pale yellow-cream | alum | decomposition |
| | cream | none | decomposition |
| ***Psoralidium tenuiflorum*** SLIM-FLOWER SCURFPEA | | | |
| flowers | pale yellow-cream | alum | heat |
| | cream | none | heat |
| ***Ptilimnium nuttallii*** MOCK BISHOP'S-WEED | | | |
| roots | cream | none | heat |
| ***Pyrrhopappus grandiflorus*** MORNING STAR | | | |
| leaves | cream | none | heat |
| | cream | none | decomposition |
| ***Ratibida columnifera*** MEXICAN HAT | | | |
| leaves and stems | pale yellow-cream | alum | decomposition |
| | cream | none | heat |
| | cream | none | decomposition |
| ray florets | cream | alum | heat |

| DYE SOURCE | DYE RESULT | MORDANT | PROCESSING |
|---|---|---|---|
| ***Ratibida columnifera,*** ray florets, continued | | | |
| | cream | copper | heat |
| | cream | none | heat |
| roots | pale yellow-cream | alum | heat |
| | cream | none | heat |
| ***Rhus copallinum*** WINGED SUMAC | | | |
| bark | pale yellow-cream | none | decomposition |
| ***Rhus glabra*** SMOOTH SUMAC | | | |
| bark | pale yellow-cream | alum | decomposition |
| | pale yellow-cream | none | decomposition |
| ***Robinia hispida*** BRISTLY LOCUST | | | |
| flowers | cream | copper | decomposition |
| | cream | none | decomposition |
| stems | cream | alum | decomposition |
| | cream | copper | decomposition |
| | cream | iron | decomposition |
| | cream | tin | decomposition |
| | cream | none | decomposition |
| ***Robinia pseudoacacia*** BLACK LOCUST | | | |
| bark | cream | alum | decomposition |
| | pale yellow-cream | copper | decomposition |
| | pale yellow-cream | iron | decomposition |
| | pale yellow-cream | none | heat |
| | cream | none | decomposition |
| leaves | pink-cream | none | decomposition |
| petals | cream | alum | decomposition |
| | pale yellow-cream | copper | decomposition |
| | pale yellow-cream | iron | decomposition |
| | cream | none | heat |
| | cream | none | decomposition |
| ***Rudbeckia grandiflora*** ROUGH CONEFLOWER | | | |
| leaves and stems | pale yellow-cream | alum | heat |
| | pale yellow-cream | none | heat |
| ray florets | pale yellow-cream | alum | heat |
| | pale yellow-cream | alum | decomposition |
| | pale yellow-cream | none | heat |
| rhizomes | pale yellow-cream | alum | heat |
| | cream | alum | decomposition |
| | cream | none | heat |
| ***Rudbeckia hirta*** BLACK-EYED SUSAN | | | |
| leaves | pale yellow-cream | alum | decomposition |
| | cream | none | decomposition |
| ray florets | pale yellow-cream | alum | decomposition |
| | pale yellow-cream | none | heat |
| | pale yellow-cream | none | decomposition |
| roots | pale yellow-cream | alum | heat |
| | cream | alum | decomposition |
| | pale yellow-cream | iron | decomposition |
| | cream | none | heat |
| | cream | none | decomposition |

| DYE SOURCE | DYE RESULT | MORDANT | PROCESSING |
|---|---|---|---|
| ***Rumex altissimus*** PALE DOCK | | | |
| flowers | cream | alum | decomposition |
| | cream | tin | decomposition |
| | cream | none | decomposition |
| leaves | cream | alum | decomposition |
| | cream | copper | decomposition |
| | pale yellow-cream | none | heat |
| | cream | none | decomposition |
| ***Sabatia campestris*** PRAIRIE ROSE GENTIAN | | | |
| petals | pale yellow-cream | iron | decomposition |
| | cream | none | heat |
| leaves and stems | pale yellow-cream | iron | decomposition |
| ***Salix caroliniana*** CAROLINA WILLOW | | | |
| leaves | pale yellow-cream | none | heat |
| | cream | none | decomposition |
| ***Salix exigua*** SANDBAR WILLOW | | | |
| stems | cream | none | heat |
| ***Salix nigra*** BLACK WILLOW | | | |
| bark | pale yellow-cream | alum | decomposition |
| | pale yellow-cream | iron | decomposition |
| | pale yellow-cream | tin | decomposition |
| | pale yellow-cream | none | decomposition |
| branchlets | cream | alum | decomposition |
| | pale yellow-cream | tin | decomposition |
| | cream | none | decomposition |
| leaves | pale yellow-cream | iron | decomposition |
| | pale yellow-cream | none | heat |
| | cream | none | decomposition |
| ***Sambucus canadensis*** ELDERBERRY | | | |
| bark | pale yellow-cream | tin | decomposition |
| leaves | pale yellow-cream | none | decomposition |
| ***Sanguinaria canadensis*** BLOODROOT | | | |
| leaves and stems | pale yellow-cream | alum | heat |
| | cream | none | heat |
| | pale yellow-cream | none | decomposition |
| ***Sapindus drummondii*** SOAPBERRY | | | |
| bark | pale yellow-cream | alum | heat |
| | pale yellow-cream | copper | heat |
| | cream | none | heat |
| | cream | none | decomposition |
| flowers | pale yellow-cream | none | heat |
| fruits | cream | alum | heat |
| | cream | iron | heat |
| | cream | none | heat |
| leaves | cream | none | decomposition |
| stems | cream | none | decomposition |
| ***Sassafras albidum*** SASSAFRAS | | | |
| flowers | cream | none | decomposition |
| leaves | pale yellow-cream | none | decomposition |

| DYE SOURCE | DYE RESULT | MORDANT | PROCESSING |
|---|---|---|---|
| ***Sassafras albidum,*** continued | | | |
| roots | pale yellow-cream | alum | decomposition |
| | cream | none | decomposition |
| ***Sedum nuttallianum*** YELLOW STONECROP | | | |
| fruits, flowers, and stems | pale yellow-cream | alum | heat |
| | pale yellow-cream | none | heat |
| ***Silene stellata*** STARRY CAMPION | | | |
| flowers without sepals | cream | alum | heat |
| | cream | copper | heat |
| | cream | iron | heat |
| | cream | none | heat |
| leaves and upper stems | cream | alum | heat |
| | cream | alum | decomposition |
| | cream | copper | heat |
| | cream | copper | decomposition |
| | cream | iron | heat |
| | cream | iron | decomposition |
| | cream | tin | decomposition |
| | cream | none | heat |
| | cream | none | decomposition |
| stems (basal) and rootstocks | cream | alum | heat |
| | cream | none | heat |
| ***Silphium integrifolium*** PRAIRIE ROSINWEED | | | |
| disk and ray florets | cream | alum | heat |
| | cream | alum | decomposition |
| | pale yellow-cream | copper | decomposition |
| | cream | iron | decomposition |
| | cream | none | heat |
| leaves | pale yellow-cream | tin | decomposition |
| | cream | none | decomposition |
| ***Sisyrinchium angustifolium*** BLUE-EYED GRASS | | | |
| leaves | pale yellow-cream | none | heat |
| ***Smilax bona-nox*** GREENBRIER | | | |
| leaves and stems | pale yellow-cream | alum | heat |
| | cream | none | decomposition |
| samaras | pale yellow-cream | tin | decomposition |
| ***Solanum dimidiatum*** WESTERN HORSE-NETTLE | | | |
| petals | cream | alum | heat |
| | cream | none | heat |
| | pale yellow-cream | none | decomposition |
| roots | cream | alum | heat |
| | cream | none | heat |
| stamens | cream | alum | heat |
| | cream | none | heat |
| ***Solanum elaeagnifolium*** SILVERLEAF NIGHTSHADE | | | |
| petals | cream | alum | heat |
| | cream | none | heat |
| roots | cream | alum | heat |
| | cream | none | heat |

| DYE SOURCE | DYE RESULT | MORDANT | PROCESSING |
|---|---|---|---|
| ***Solidago missouriensis*** PLAINS GOLDENROD | | | |
| rhizomes and roots | pale yellow-cream | alum | decomposition |
| | cream | none | heat |
| | cream | none | decomposition |
| ***Solidago rigida*** STIFF GOLDENROD | | | |
| roots | cream | alum | decomposition |
| | cream | tin | decomposition |
| | cream | none | heat |
| ***Sorghastrum nutans*** INDIANGRASS | | | |
| leaves | pale yellow-cream | alum | decomposition |
| | cream | none | heat |
| | pale yellow-cream | none | decomposition |
| rhizomes and roots | pale yellow-cream | alum | heat |
| | pale yellow-cream | iron | heat |
| | cream | none | heat |
| spikelets | cream | none | heat |
| | cream | none | decomposition |
| stems | cream | alum | heat |
| | cream | alum | decomposition |
| | pale yellow-cream | iron | heat |
| | cream | iron | decomposition |
| | pale yellow-cream | tin | decomposition |
| | cream | none | heat |
| | cream | none | decomposition |
| ***Stillingia sylvatica*** QUEEN'S DELIGHT | | | |
| roots | cream | none | decomposition |
| stems | pale yellow-cream | alum | decomposition |
| | pale yellow-cream | copper | decomposition |
| | cream | none | decomposition |
| ***Streptanthus hyacinthoides*** SMOOTH TWIST-FLOWER | | | |
| flowers | pale yellow-cream | alum | heat |
| | pale yellow-cream | none | heat |
| ***Tephrosia virginiana*** GOAT'S RUE | | | |
| fruits and bracts | cream | none | decomposition |
| leaves | pale yellow-cream | none | decomposition |
| rhizomes | pale yellow-cream | alum | decomposition |
| | cream | tin | decomposition |
| | pale yellow-cream | none | decomposition |
| roots | cream | none | decomposition |
| ***Teucrium canadense*** AMERICAN GERMANDER | | | |
| flowers | pale yellow-cream | alum | heat |
| | cream | none | heat |
| roots | cream | alum | decomposition |
| | cream | copper | decomposition |
| | pale yellow-cream | iron | decomposition |
| | cream | tin | decomposition |
| | cream | none | heat |
| | cream | none | decomposition |

| DYE SOURCE | DYE RESULT | MORDANT | PROCESSING |
|---|---|---|---|
| ***Tradescantia occidentalis*** PRAIRIE SPIDERWORT | | | |
| stems | pale yellow-cream | alum | heat |
| | pale yellow-cream | iron | decomposition |
| | cream | none | heat |
| ***Ulmus rubra*** SLIPPERY ELM | | | |
| bark | pale yellow-cream | tin | decomposition |
| ***Vernonia baldwinii*** WESTERN IRONWEED | | | |
| stems | pale yellow-cream | tin | decomposition |
| | cream | none | heat |
| ***Viburnum rufidulum*** RUSTY BLACK HAW | | | |
| flowers | pale yellow-cream | alum | decomposition |
| | cream | none | heat |
| leaves | cream | alum | heat |
| | pale yellow-cream | alum | decomposition |
| | cream | none | heat |
| pedicels | pale yellow-cream | alum | heat |
| | cream | none | heat |
| roots | pale yellow-cream | tin | decomposition |
| | cream | none | decomposition |
| ***Vitis aestivalis*** PIGEON GRAPE | | | |
| bark | pale yellow-cream | alum | decomposition |
| | pale yellow-cream | iron | decomposition |
| | pale yellow-cream | none | decomposition |
| leaves | pale yellow-cream | alum | decomposition |
| | pale yellow-cream | none | heat |
| | cream | none | decomposition |

# 12

# A Catalog of Native Dye Plants

This chapter describes the plants that are listed as dye sources in chapters 5 through 10. Plants are arranged alphabetically by the scientific name of each species, followed by the common name or names and the dye colors produced by various parts of the plant. Each entry includes a description of the plant's morphology, biology, soil requirements, and distribution, and ethnobotanical information.

The information presented is a compilation of our observations of the species in the field, our examination of herbarium specimens, and information appearing in taxonomic manuals, books, and journal articles. Unless otherwise indicated, information on the morphology and ecology of these plants is compiled primarily from Fernald (1950), Correll and Johnston (1970), Great Plains Flora Association (1986), Gleason and Cronquist (1991), and Flora of North America Editorial Committee (1993–2003); information on their Native American ethnobotany is compiled from Weiner (1972), Krochmal and Krochmal (1973), Kindscher (1992), and Moerman (1986, 1998); toxicity information is taken from Millsbaugh (1892), Mitchell and Rook (1979), Morton (1982), and Burrows and Tyrl (2001); and information on wildlife uses is collected from Martin, Zim, and Nelson (1951), Stubbendieck, Hatch, and Butterfield (1992), Hatch and Pluhar (1993), and Miller and Miller (1999).

In some instances "new" scientific names are used for familiar species. These names reflect changes in classification, resulting from systematic studies that have provided additional information and a better understanding of relationships among taxa. In a few instances the rules of botanical nomenclature mandate these changes. In general we employ the nomenclature and classification used by the Flora of North America Editorial Committee (1993–2003). Author citations (the name or abbreviation of the name of the person or people who published the species name) are taken from Brummitt and Powell (1992) and the Plant Names Project (1999). The common names cited are those most frequently used based on our experience, their citation in floristic works, and standardized lists such as the U.S. Department of Agriculture's *PLANTS Database* (2002).

## *Acer negundo* L.

BOXELDER, ASH-LEAVED MAPLE

**brown, green, orange, yellow**

A member of the Aceraceae (maple family), *Acer negundo* is a small to medium tree 10–20 m tall that is readily recognized by its opposite, pinnately compound leaves with three or five yellow-green leaflets, its dark green branchlets, and its clusters of double samaras. Unlike other maples, it has compound rather than simple leaves, which in autumn turn bright yellow. Typically encountered in bottomlands or along stream margins, it is the most widely distributed species of the genus in North America and occurs across the continent. It is the only member of the genus in the prairie regions. Native Americans used the straight branchlets for arrow shafts, made

Leaves of *Acer negundo*.

decoctions of the inner bark for use as an emetic, and made sugar from the sap. Small mammals and birds eat the fruits, and birds use the leaves and pedicels in nest building.

## *Acer saccharinum* L.

SILVER MAPLE, SOFT MAPLE

**black, brown, red, yellow**

A member of the Aceraceae (maple family), *Acer saccharinum* is a medium tree up to 30 m tall, with opposite, palmately lobed leaves that are light green above and silvery white below. The lobes are deep and coarsely toothed. The light gray bark of the trunk splits into long plates. The flowers appear before the leaves and produce large double samaras with wings that diverge at almost right angles to each other. This species occurs throughout the eastern half of the continent and is characteristic of moist or wet soils, especially along river and creek banks. Although widely planted as an ornamental or shade tree because of its fast growth and yellow to yellow-red autumn foliage, *A. saccharinum* is prone to break limbs in high winds and to cause uplifting of sidewalks and patios because of its vigorous root growth. Native Americans made infusions and decoctions from the species to treat a variety of ailments including diarrhea, cramps, measles, and venereal diseases. Tribes in northern states made sugar from the sap. Small mammals and birds eat the fruits, and birds use the leaves and pedicels in nest building.

Leaves of *Acer saccharinum*.

## *Achillea millefolium* L.

YARROW

**black, green, yellow**

A member of the Asteraceae (sunflower family), *Achillea millefolium* is an aromatic, perennial herb 20–70 cm tall that is readily recognized by its fern-like leaves and flat-topped or hemispheric clusters of small heads with

Heads of *Achillea millefolium*. Photo by Steven K. Goldsmith.

white ray florets. It is sometimes confused with *Daucus carota* (Queen Anne's lace), but close examination of the inflorescence reveals heads rather than individual flowers. When the leaves are crushed, the odor of menthol is distinctive. Morphologically quite variable and forming distinct ecotypes, the species occurs across the continent. The binomial *A. lanulosa* appears in many older floristic treatments. Characteristic of the mid to late stages of plant succession, *A. millefolium* occupies a variety of soil types and habitats. Plants are present in prairies, meadows, open woods, and slightly disturbed sites. They increase in abundance in pastures that are heavily grazed. Flowering occurs in late spring and early summer. The genus name is derived from Achilles, who supposedly used the plant to heal the wounds of his soldiers. During the Civil War, soldiers did indeed apply the crushed leaves to their wounds. Historical common names—staunchweed, knight's milfoil, bloodwort, woundwort—all allude to this use. Some sixty Native American tribes used the species to treat burns, earaches, coughs, throat irritations, colds, nausea, and bleeding. Salves and teas were also made. The dried inflorescences are sometimes dyed and used in floral arrangements.

## *Ageratina altissima* (L.) R.M. King & H. Rob.

WHITE SNAKEROOT, WHITE SANICLE, FALL POISON, WHITETOP, POOLWORT

**black, brown, green, yellow**

A member of the Asteraceae (sunflower family), *Ageratina altissima* is a perennial herb 30–150 cm tall from fibrous rooted crowns or short rhizomes. It is readily recognized by its opposite, ovate, rugose, three-ribbed leaves and its showy clusters of small white heads bearing only disk florets. Crowned by a pappus of capillary bristles, the mature achenes are cylindrical, five-ribbed, and black. Characteristic of the mid to late stages of plant succession, this species is typically found in moist, rich soils of shaded sites such as wooded ravines, stream banks and terraces, and sheltered slopes. Large dense stands may develop, and it may become the dominant understory species in the area. It occurs throughout the eastern half of the continent as far west as eastern Nebraska, Kansas, Oklahoma, and Texas. Growth begins in May or June, with flowering from August through October. The leaves and stems are toxic and cause a disease known as trembles in cattle and milk sickness in humans. These diseases are of considerable historical interest because of the suffering and panic they caused among early-nineteenth-century settlers in the Mississippi and Ohio river valleys. In some areas of Ohio and Indiana, one-quarter to one-half of the deaths in the early 1800s were attributed to milk sickness. This species was long known as *Eupatorium rugosum*.

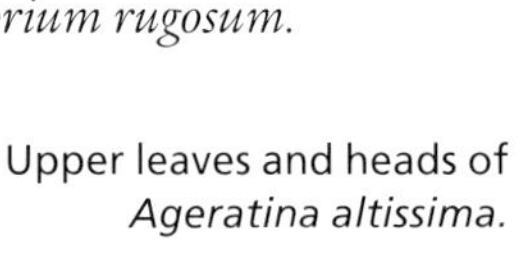

Upper leaves and heads of *Ageratina altissima*.

## ***Alnus maritima*** (Marshall) Muhl. ex Nutt.

SEASIDE ALDER

**black, brown, green, orange, yellow**

A member of the Betulaceae (birch family), *Alnus maritima* is a bushy shrub up to 4 m tall that bears alternate, simple, oblong to obovate, serrate leaves. Flowering in the late summer or spring, it produces pendulous, usually clustered staminate catkins and short, woody, ovoid to ellipsoid, pistillate catkins that

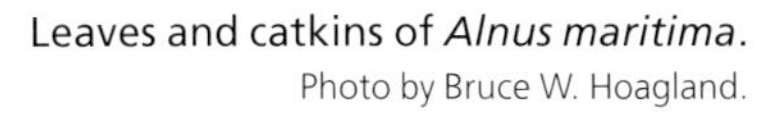

Leaves and catkins of *Alnus maritima*.
Photo by Bruce W. Hoagland.

persist on the branches. Occurring at the edges of streams and ponds, often in standing water, this species has an unusual geographic distribution. It is known only in the Delmarva Peninsula of Delaware and Maryland, in Georgia, and in the Blue River and Pennington Creek drainages of south-central Oklahoma. Taxonomists continue to debate whether populations in the three areas are relics of a once more widely distributed species or whether their distribution is due to human activities. For example, Native Americans may have introduced plants into Oklahoma when they were relocated there from the southeastern United States in the 1830s. Specific use of *A. maritima* by Native Americans is not cited in the ethnobotanical literature, but its relatives were used as sources of medicine, dye, fiber, and fuel for smoking meat.

### *Ambrosia psilostachya* DC.

WESTERN RAGWEED

**black, brown, green, yellow**

A member of the Asteraceae (sunflower family), *Ambrosia psilostachya* is a perennial herb 30–60 cm tall that often forms extensive clones from woody, rhizome-like roots. The pinnatifid, gray-green, subsessile leaves are opposite at the lower nodes and alternate at the upper. Inconspicuous and imperfect, the flowers are borne either in staminate heads arranged in terminal racemes or pistillate heads forming hard burs in the upper leaf axils. The species occurs across the continent, except for the Rocky Mountains and the east-central states. It is characteristic of the mid to late stages of plant succession and is always present in prairies, but it increases with heavy grazing. *Ambrosia psilostachya* is regarded as a classic indicator of land use; the greater its abundance, the greater the disturbance of the site, especially when due to heavy grazing. Native Americans brewed its leaves for a tea to relieve intestinal cramps, diarrhea, and sore eyes. However, like its relatives, its pollen causes severe hay fever in many humans, and can even be dangerous to especially sensitive individuals.

Flowering plant of *Ambrosia psilostachya*.

## *Ambrosia trifida* L.

GIANT RAGWEED

**black, brown, green, yellow**

A member of the Asteraceae (sunflower family), *Ambrosia trifida* is an annual herb whose sap typically stains skin and clothes purple-red. Arising from large taproots, its stems are 1–4 m tall, angular, striate, and scabrous, bearing opposite, long-petioled, dark green leaves with palmately three-, five-, or seven-lobed blades. Its imperfect flowers are borne in terminal racemes of nodding, hemispheric, staminate heads or in axillary clusters of hard burs of pistillate heads. Taxonomists believe this species to be native to the central portion of the continent from Canada to northern Mexico and adventive in the far west and east. In contrast, agronomists consider it to be native to Europe. Often dominating the early stages of plant succession, it is regarded as a classic indicator of disturbed soils: the greater the disturbance, the greater its abundance. Archaeological studies of prehistoric North American campsites indicate that early humans ate its achenes. Plains tribes and European settlers brewed the leaves for teas to relieve intestinal cramps, diarrhea, and sore eyes. Despite its generic name, which means "nectar of the gods," giant ragweed is perhaps the most notorious plant known to cause hay fever. August 15 to October 1 is often cited as hay fever season, a period that coincides with the species' flowering. Sensitive individuals may react to as few as one hundred pollen grains of *A. trifida* per cubic meter of air and suffer severe and dangerous respiratory problems.

Leaves and staminate heads of *Ambrosia trifida*. Photo by George M. Diggs Jr.

## *Amorpha canescens* Pursh

LEADPLANT

**brown, green, yellow**

A member of the Fabaceae (pea family), *Amorpha canescens* is a perennial herb that is sometimes woody at the base. It is readily recognized by its silver-gray foliage; its sessile or subsessile, crowded, pinnately compound

Flowering plant of *Amorpha canescens*. Photo by Charles S. Lewallen.

leaves with thirteen to twenty-four leaflets; and its elongate, spicate, bright purple or violet racemes arising from the uppermost nodes. The species occurs in the Midwest from Ontario and Manitoba south to central Texas and northeastern New Mexico. Characteristic of the late stages of plant succession in tallgrass prairies, *A. canescens* also occurs in glades and open woodlands. Flowering occurs from May to July. This species is considered one of the most important prairie legumes. Its foliage, which is high in nutrient content, is readily eaten by livestock and occasionally eaten by white-tailed deer. Its abundance can be used to assess the seral stage of a range. Native Americans used plants as a beverage, for smoking, and as a treatment for eczema, pinworms, stomach pain, rheumatism, and neuralgia.

## *Apocynum cannabinum* L.

INDIAN HEMP, PRAIRIE DOGBANE, HEMP DOGBANE

**brown, green, orange, yellow**

A member of the Apocynaceae (dogbane family), *Apocynum cannabinum* is a rhizomatous, perennial herb 20–100 cm tall, with erect, glabrous to villous stems that branch above their middles and bear opposite, ascending to erect leaves. Its sap is white and viscous. Small white to greenish flowers are borne in dense terminal cymes. Most conspicuous are the pairs of long, slightly curved, cylindrical, pendulous follicles typically joined at the stigmas and bearing small seeds with tufts of long silky hairs. Although it occurs across the continent, this species is most commonly encountered in moist soils of ditches, along waterways, and in open woodlands and fields. It often forms dense stands of plants. As is implied by the taxon's common name, the mature stems are tough and fibrous. They were used by Native Americans to make fishing nets and lines, and for twine used in basket making. Like other members of the family, *A. cannabinum* contains cardiotoxic glycosides, but its hazard is negligible. Besides being very fibrous, the foliage is apparently quite distasteful and is generally ignored by livestock. In humans, only a few instances of adverse effects have been reported.

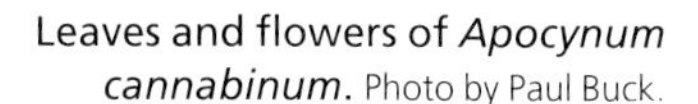

Leaves and flowers of *Apocynum cannabinum*. Photo by Paul Buck.

## ***Argemone polyanthemos*** (Fedde) G.B. Ownbey

PRICKLY POPPY

**brown, green, orange, yellow**

A member of the Papaveraceae (poppy family), *Argemone polyanthemos* is an annual or biennial herb arising from a deep taproot and producing a viscous, bright yellow sap. Its simple, alternate leaves are conspicuously prickly, as are its sepals and capsules. The large showy flowers bear six white petals and 150 or more stamens. Usually encountered in sandy soils of dunes, prairies, floodplains, and roadsides or other disturbed sites, this species is distributed primarily in the central prairies and plains from South Dakota and eastern Wyoming to northern Texas and eastern New Mexico. As is characteristic of the family, *A. polyanthemos* contains isoquinoline alkaloids, which affect a variety of body systems. The genus name is derived from the Latin root *argema*, meaning "cataract of the eye." Native Americans used the sap and seeds to treat eye problems. Crushed

Flower of *Argemone polyanthemos*.

seeds were used in salves for burns and emetics. The yellow dye obtained from this plant was used by Kiowas for tattooing and by Lakotas for decorating arrows.

### *Asclepias arenaria* Torr.

SAND MILKWEED

**black, brown, yellow**

A member of the Asclepiadaceae (milkweed family), *Asclepias arenaria* is a perennial herb that arises from a deep rhizome and produces a milky sap. Its stems are 20–50 cm tall and bear large, simple, opposite leaves with broadly obovate to oval blades and truncate to cordate apices. Its small, pale green flowers are borne in umbellate cymes in the axils of the upper leaves and produce large spindle-shaped follicles bearing numerous seeds with tufts of hairs. These flowers feature a gynostegium, as is characteristic of the family. As its common name implies, this species occurs primarily in sandy soils of prairies and along roadsides. A central plains and prairie species, it is distributed from eastern Wyoming and southern South Dakota south to Texas, New Mexico, and northern Chihuahua, Mexico. The genus is well known for the cardiotoxic and neurotoxic alkaloids that it produces.

Flowering plant of *Asclepias tuberosa*.

### *Asclepias tuberosa* L.

BUTTERFLY MILKWEED, ORANGE MILKWEED, CHIGGER-FLOWER

**green, yellow**

A member of the Asclepiadaceae (milkweed family), *Asclepias tuberosa* is a perennial herb that arises from a deep woody rootstock. Unlike the other members of the genus, its sap is not milky. Multiple stems 50–90 cm tall bear numerous simple, alternate leaves with lanceolate to oblong, dark green blades. Its small yellow to orange or red flowers are borne in showy, terminal or subterminal, umbellate cymes. The large spindle-shaped follicles bear numerous seeds with tufts of hairs. Flowers feature a gynostegium, as is characteristic of the family. Occurring in sandy, loamy, or calcareous soils of prairies and open woods, this species is distributed across the United States except for the northwest portion. As the name "butterfly milkweed" indicates, it attracts a plethora of butterflies and other insects. It is cultivated as a garden ornamental. Native Americans and early settlers used the roots of *A. tuberosa*, at one time called pleurisy-root, to treat lung inflammation, and as an analgesic, laxative, dermatological aid, and gynecological aid.

### ***Asclepias viridis*** Walter

ANTELOPE-HORN MILKWEED, GREEN ANTELOPE-HORN, ANTELOPE-HORN

**brown, green, yellow**

A member of the Asclepiadaceae (milkweed family), *Asclepias viridis* is a perennial herb that arises from a thickened vertical rootstock and produces a milky sap. Its stems are solitary or clustered, ascending or decumbent, and 25–70 cm tall. Oriented at right angles to each other, the pairs of opposite, simple leaves are oblong or ovate to lanceolate. The solitary, showy inflorescences are large, spherical, and bear pale green and purple flowers. Ribbed, spindle-shaped follicles bearing numerous seeds with tufts of hairs appear at the end of flowering. The flowers feature a gynostegium, as is characteristic of the family. Generally occupying clayey or loamy soils and characteristic of the mid stages of plant succession, *A. viridis* typically occurs as either scattered plants or dense populations in open prairies, dry open woods, heavily grazed pastures, and disturbed sites. It occurs in the United States from West Virginia, Ohio, the Carolinas, and Florida west to Nebraska, Kansas, Oklahoma, and western Texas. Flowering typically occurs in May and July, although plants occasionally flower in late summer or early fall. The genus is well known for the cardiotoxic and neurotoxic alkaloids that it produces. The binomial *Asclepiodora viridis* is encountered in older publications.

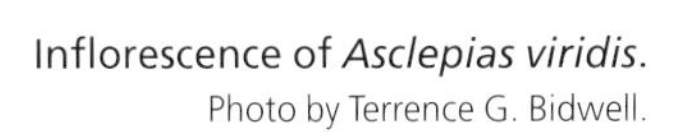

Inflorescence of *Asclepias viridis*.
Photo by Terrence G. Bidwell.

### *Asimina triloba* (L.) Dunal

PAWPAW, POOR MAN'S BANANA, FALSE BANANA, AMERICAN CUSTARD-APPLE

**black, brown, green, red, yellow**

A member of the Annonaceae (custard-apple family), *Asimina triloba* is a deciduous large shrub or small tree up to 10 m tall. It bears alternate, simple, oblong to obcordate, acuminate leaves 15–35 cm long that are aromatic when crushed and that turn a brilliant yellow in autumn. Solitary flowers arise from the wood of the previous year and bear six petals that are initially green, then brown, and finally lurid purple at maturity. The fruit is a berry with several large flattened seeds. Characteristically present as an understory species in rich, moist soils of deciduous forests, this species is distributed throughout the eastern half of the continent. The berry is edible and similar in flavor to a banana, as the common names imply. Eaten only after the first frost, the custard-like flesh is characterized by some people as delicious and by others as insipid. Upon handling or eating the berries, some individuals experience dermatitis or irritation of the digestive tract. At one time a tincture made from the seeds was used as an emetic. Cherokees used the inner bark to make string and rope.

Leaves and flower of *Asimina triloba*. Photo by Paul Buck.

### *Baccharis salicina* Torr. & A. Gray

WILLOW BACCHARIS, FALSE WILLOW, GREAT PLAINS FALSE WILLOW

**black, brown, green, yellow**

A member of the Asteraceae (sunflower family), *Baccharis salicina* is a dioecious shrub 1–3 m tall, with alternate, simple, sessile or subsessile leaves. Its blades are oblanceolate-oblong and serrate. The flowers are borne in paniculate inflorescences of numerous small staminate and pistillate heads, and when in fruit give the plants a fluffy or feathery appearance. Occurring in open,

Heads of *Baccharis salicina*. Photo by Bruce W. Hoagland.

sandy, often weakly saline floodplains, this species is distributed from Texas and New Mexico north to Kansas, Colorado, and southeastern Utah.

### ***Baptisia australis*** (L.) R. Br.

BLUE WILD-INDIGO, BLUE FALSE-INDIGO

**black, brown, green, yellow**

A member of the Fabaceae (pea family), *Baptisia australis* is a perennial herb that is readily recognized by its stout, erect stems that are unbranched below and branched above; its alternate, palmately compound, sessile or subsessile leaves with three obovate or oblanceolate leaflets; its elongate racemes of showy flowers with blue to purple, papilionaceous corollas; and its large, sharp-pointed, plump legumes. The dark gray to black appearance of the foliage after fruiting is characteristic. Occurring as either scattered plants or small localized populations, this species is characteristic of the late stages of plant succession, in prairies, limestone glades, and open woods. It is distributed throughout the eastern half of the United States and extends west to southeastern Nebraska, central Kansas, central Oklahoma, and north-central Texas. Flowering is from April to June. Native Americans used the fruits for rattles and used various parts of the plant as a source of blue dye, a purgative, an emetic, and a toothache remedy. The species hybridizes with *B. bracteata*. Synonyms appearing in the older literature are *B. vespertina*, *B. minor*, and *B. texana*.

Flowering plant of *Baptisia australis*.

### ***Baptisia bracteata*** Muhl. ex Elliott

PLAINS WILD-INDIGO, LARGE-BRACTED WILD-INDIGO, YELLOW FALSE-INDIGO, LONG-BRACT WILD-INDIGO

**brown, green, yellow**

A member of the Fabaceae (pea family), *Baptisia bracteata* is a perennial herb arising from a massive rootstock. It is readily recognized by its alternate, palmately compound leaves with three oblanceolate leaflets. Two large stipules sometimes give it the appearance of having five leaflets. Also charac-

Flowering plant of *Baptisia bracteata*.

teristic are its elongate, drooping racemes of showy flowers with their yellowish or cream-white, papilionaceous corollas and its plump, reticulately veined legumes with their elongate beaks and stipes. Occurring either as scattered plants or small localized populations, this species is characteristic of the late stages of plant succession. Plants occupy sandy, loamy, and clayey soils in prairies and open woods. They often occur on roadsides and on thin, bare soils of rocky sites. Widely distributed throughout the eastern half of the United States, this species extends west to southeastern Nebraska, central Kansas, central Oklahoma, and north-central Texas. Flowering is from April to June. Native Americans made infusions from the roots, decoctions from the leaves, and salves from the seeds for various ailments. The fruits were used for rattles. The binomial *B. leucophaea* has long been used for this species. Taxonomic studies reveal, however, that it is conspecific with *B. bracteata* of the southeastern coastal plain, which has nomenclatural priority and must be used.

### *Baptisia sphaerocarpa* Nutt.

GOLDEN WILD-INDIGO

**brown, yellow**

A member of the Fabaceae (pea family), *Baptisia sphaerocarpa* is a perennial herb up to 1 m tall, with several branching stems arising from a large root-

Flowering plant of *Baptisia sphaerocarpa*. Photo by George M. Diggs Jr., courtesy of the Botanical Research Institute of Texas.

stock. It is readily recognized by its alternate, palmately compound leaves with three broadly oblanceolate to obovate leaflets. The upper leaves typically have only one or two leaflets. Also characteristic are its terminal racemes, bearing numerous bright flowers with golden yellow papilionaceous corollas, and its small, almost spherical, brown legumes that are stipitate and slender-beaked. Generally occurring as small, dense, localized populations, this species is found in sandy or loamy soils of prairies and pastures in Oklahoma, Arkansas, Texas, and Louisiana. *Baptisia sphaerocarpa* hybridizes with both *B. australis* and *B. bracteata* to produce showy intermediate individuals that have been formally recognized as *B. ×variicolor* and *B. ×bushii* (Kosnik et al. 1996).

## ***Bidens aristosa*** (Michx.) Britton

BEARDED BEGGAR-TICKS, AWNLESS BEGGAR-TICKS, TICKSEED SUNFLOWER

**brown, orange, yellow**

A member of the Asteraceae (sunflower family), *Bidens aristosa* is an annual or biennial herb with erect stems 30–100 cm tall that bear opposite, petiolate, one- or two-pinnately dissected leaves. Its flowers are borne in numerous funnelform heads. The yellow ray florets typically number eight and are showy. The yellow disk florets produce flattened brown or nearly black achenes that may bear two barbed awns or merely short teeth. As presently circumscribed, this species includes plants formerly known as *B. polylepis*. Though distributed throughout the eastern half of the United States, this species is most commonly encountered in the Midwest and Great Plains. Plants occur in moist sites such as borrow ditches, stream banks, and marshes. Flowering is from August to October, and when populations are dense in borrow ditches, the species is quite conspicuous and attractive.

Heads of *Bidens aristosa*. Photo courtesy of the Samuel Roberts Noble Foundation.

### *Bidens bipinnata* L.

SPANISH NEEDLES, BEGGAR-TICKS, BEGGAR-LICE, PITCHFORKS

**black, brown, green, orange, yellow**

A member of the Asteraceae (sunflower family), *Bidens bipinnata* is an annual herb with erect, square stems 30–170 cm tall that bear opposite (uppermost sometimes alternate), two- or three-pinnately dissected leaves. The flowers are borne in numerous narrowly oblong heads, with the yellowish white ray florets being inconspicuous and not exceeding the yellow disk florets. Its stoutly awned, four-angular, black achenes are readily embedded in skin and clothing. This species is distributed across the eastern half of the continent. Characteristic of the early stages of plant succession, it is typically encountered in moist sites, especially in disturbed areas. Flowering is from July to October. The achenes persist on the plants from August to February, and their dispersal by wildlife is facilitated by the retrorse barbs on the awns. Cherokees made an infusion with this plant to treat worms and chewed the leaves to treat sore throats.

Head of *Bidens bipinnata*. Photo by Dan Tenaglia.

### *Brickellia eupatorioides* (L.) Shinners

FALSE BONESET

**black, brown, green, yellow**

A member of the Asteraceae (sunflower family), *Brickellia eupatorioides* is a multistemmed, perennial herb that grows 30–100 cm tall from a woody, branching rootstock. Its alternate, simple, lanceolate, serrate to entire leaves are gland-dotted beneath. Borne in corymbose clusters, the numerous funnelform heads contain only creamy or yellowish white disk florets. In fruit, the columnar ten-ribbed achenes have a light brown or tan plumose pappus. Occurring primarily in loose sandy soils of prairies and plains, this species is distributed throughout the eastern half of the United States. Navajos made decoctions of the roots to treat coughs and old injuries. False boneset was originally classified in the genus *Kuhnia*; however, systematic work indicates that the two genera should be united.

## *Buchnera americana* L.

BLUEHEARTS

**black, brown, green**

A member of the Scrophulariaceae (figwort family), *Buchnera americana* is a perennial, semiparasitic herb with solitary, erect, unbranched stems 30–70 cm tall. Its leaves are opposite, simple, and progressively smaller along the stem. Borne in the axils of small bracts in terminal spikes, the salverform violet to blue or purple flowers are almost radially symmetrical, in contrast with the bilabiate flowers typical of the family. In fruit, the capsules are ovoid-oblong and usually gibbous at the base. Although photosynthetic, *B. americana* does obtain some nutrients by parasitizing the roots of adjacent grasses. The entire plant turns black upon drying.

Flowers of *Buchnera americana*.
Photo by Adam K. Ryburn.

## *Bumelia lanuginosa* (Michx.) Pers.

CHITTAMWOOD, GUM BUMELIA, WOOLLY BUCKTHORN, CHITTAM, GUM ELASTIC, GUM BULLY

**black, brown, green, yellow**

A member of the Sapotaceae (sapodilla family), *Bumelia lanuginosa* is a large shrub or medium tree 12–18 m tall. It occasionally forms thickets in sandy soils. Its bark is gray-brown tinged with red and deeply fissured, typically in a rhomboidal pattern. Often terminating in stout thorns, the branchlets bear alternate, simple, oblanceolate leaves fascicled on short spurs. The blades are shiny and dark green above, and rusty-brown or gray-green and woolly-villous below. The numerous greenish or cream-white flowers are borne in cymose clusters on spurs and produce obovoid purple-black berries that turn golden brown when dry. Generally encountered as solitary trees in open woods on rocky, thin soils in uplands, this species is characteristic of the mid to late stages of plant succession and occurs in a swath from east-central Missouri to Oklahoma, Texas, and northeastern Mexico. Flowering is in June and July, and bees in great numbers visit when the flowers open. Because its leaves persist, this plant provides cover in early winter for wildlife. The berries are edible, and Native Americans once used the mucilage of the outer bark as chewing gum. This plant is occasionally cultivated as an ornamental. Some taxonomists do not recognize *Bumelia* as a distinct genus and use the binomial *Sideroxylon lanuginosum* for the species.

Leaves and flowers of *Bumelia lanuginosa*. Photo courtesy of the Samuel Roberts Noble Foundation.

Leaves and verticils of fruits of *Callicarpa americana*. Photo by Ronald E. Masters.

### *Callicarpa americana* L.

AMERICAN BEAUTYBERRY, FRENCH MULBERRY, SOURBUSH, BUNCHBERRY, FOXBERRY, TURKEYBERRY, SPANISH MULBERRY

**brown, green, yellow**

A member of the Verbenaceae (vervain family), *Callicarpa americana* is a much-branched, weakly aromatic shrub 1–2 m tall, with arching stems and light gray-brown to gray, stellate-tomentose bark. The deciduous, opposite, simple leaves have serrate or crenate, elliptic to ovate blades with stellate surfaces and acuminate apices. The small flowers are borne in axillary verticils and have four or five fused bluish, pinkish, or reddish white petals. Clusters of rose-pink to red-purple drupes are quite showy when mature. Adapted to moist loamy or sandy soils, this species is generally encountered as scattered understory shrubs on wooded hillsides and in bottomlands. It also occurs in dry, rocky, shallow soils in upland sites and is characteristic of the mid stages of plant succession. Distributed in the southeastern quarter of the United States as far west as Texas, Oklahoma, and Missouri, this species flowers from June to early August, with the brightly colored drupes typically persisting on the leafless stems until January. Because of its conspicuous fruits, *C. americana* is used as an ornamental. Native Americans made decoctions of the roots, branches, and berries to treat colic, vertigo, fever, and rheumatism. It is considered a good species for wildlife and can withstand heavy use by browsers. High in moisture content and available when other fruits are scarce, the drupes are eaten by many birds and small mammals.

## *Calylophus hartwegii* (Benth.) P.H. Raven

HARTWEG'S SUNDROP, HARTWEG'S EVENING-PRIMROSE

**black, brown, orange, yellow**

A member of the Onagraceae (evening-primrose family), *Calylophus hartwegii* is a perennial herb or subshrub that arises from a woody rootstock. Its branched, decumbent or spreading stems are 5–40 cm tall and bear numerous alternate, simple leaves with linear to oblong-lanceolate, entire or dentate blades. The showy, pale yellow or pinkish orange, four-merous, radially symmetrical flowers are borne in the axils of the upper leaves. Eight stamens surround the one style that terminates in a peltate, four-sided stigma. Its capsules at the base of a tubular or slightly funnelform hypan-

Flowers of *Calylophus hartwegii*. Photo by Robert J. O'Kennon.

thium are slightly cylindrical and contain numerous seeds. Plants occur in a variety of soil types on prairie hillsides, limestone and gypsum outcrops, and along roadsides. This species is distributed in the south-central part of the United States from Kansas and Colorado south to Texas and New Mexico and west to Arizona. Flowering is from April to July. Navajos considered this plant a "life medicine" and used it to treat internal bleeding. It has long been known as *Oenothera hartwegii*; however, systematic work has resulted in division of the large genus *Oenothera*, with subsequent reclassification of it and others into the genus *Calylophus*.

### ***Calylophus serrulatus*** (Nutt.) P.H. Raven

HALFSHRUB SUNDROP, PLAINS EVENING-PRIMROSE, YELLOW SUNDROP

**black, brown, green, orange, yellow**

A member of the Onagraceae (evening-primrose family), *Calylophus serrulatus* is a perennial herb or subshrub that arises from a branched, woody rootstock. Its typically unbranched, erect or ascending stems vary from few to many, are 10–50 cm tall, and bear numerous alternate, simple leaves with linear to narrowly oblong or oblanceolate, serrulate or rarely entire blades. The showy, bright yellow, four-merous, radially symmetrical flowers are solitary in the axils of the upper leaves. Eight stamens surround the one style that terminates in a peltate, four-sided stigma. Its capsules at the bases of a cylindrical to funnelform hypanthia are slightly four-angled and contain numerous seeds. As the common name "halfshrub sundrop" implies, the plant is readily recognized by its woody stem base. Characteristic of the mid stages of plant succession and occupying a variety of soil types, it is found in dry prairies, often on eroding banks, and is distributed in the Great Plains and tallgrass prairie from Alberta, Saskatchewan, and Manitoba south to Arkansas, Texas, northeastern Mexico, and west to Arizona. Flowering is from June to September. It has long been known as *Oenothera serrulata*. Systematic work, however, has resulted in division of the large genus *Oenothera*, with subsequent reclassification of the species and others into the genus *Calylophus*. The flowers are important for some species of butterflies and moths.

Flowers of *Calylophus serrulatus*.
Photo by Terrence G. Bidwell.

Leaves and flowers of *Campsis radicans*. Photo by Robert J. O'Kennon, courtesy of the Botanical Research Institute of Texas.

### ***Campsis radicans*** (L.) Seem.

TRUMPET CREEPER, TRUMPET VINE, COWITCH VINE, TRUMPET HONEYSUCKLE

**brown, green, yellow**

A member of the Bignoniaceae (catalpa family), *Campsis radicans* is a shrubby vine that climbs 10 m or higher via aerial rootlets or sprawls across other plants or fences. Its opposite, one-pinnately compound leaves bear five to thirteen oblong-lanceolate to elliptic, serrate leaflets with long, acuminate apices. Borne in terminal inflorescences, the large, showy, orange to orange-red flowers are tubular-funnelform and slightly bilaterally symmetrical, and produce curved, spindle-shaped capsules. Native to moist woods in the eastern half of the continent, this species has been widely planted elsewhere as an ornamental. It escapes cultivation and naturalizes and is sometimes an aggressive weed in fencerows and along roadsides. The flowers are visited by hummingbirds and long-tongued bees. Sensitive individuals may develop minor dermatitis upon handling the leaves and flowers.

### ***Carya cordiformis*** (Wangenh.) K. Koch

BITTERNUT HICKORY, PIGNUT, PIGNUT HICKORY, BITTERNUT

**black, brown, orange, yellow**

A member of the Juglandaceae (walnut family), *Carya cordiformis* is a large monoecious tree up to 30 m tall. Its bark is light brown, and its slender branchlets bear alternate, one-pinnately compound leaves with seven or nine lanceolate, serrulate leaflets. The principal diagnostic characteristics for this species are the sulfurous yellow, asymmetrical, valvate terminal

buds and the nuts enclosed in involucral husks with four winged sutures that extend a third to a half of the way around the husk. Typically occurring in bottomlands and on moist slopes of uplands, this species is distributed throughout the deciduous forest of the eastern half of the continent. Its wood is used for lumber, flooring, and firewood. Unlike other hickories, it is not a good food species for wildlife, because its nuts and twigs are bitter. Iroquois used this tree extensively for food and fiber and as a diuretic, laxative, and general tonic.

Terminal bud of *Carya cordiformis*.

### *Carya illinoinensis* (Wangenh.) K. Koch

PECAN

**brown, yellow**

A member of the Juglandaceae (walnut family), *Carya illinoinensis* is a large monoecious tree up to 50 m tall. It is readily recognized by its dark green, one-pinnately compound leaves with their eleven to seventeen falcate leaflets; its heart-shaped or three-lobed leaf scars; and its brown, oblong nuts that fall from husks dehiscing along four sutures. Its three terminal leaflets are not larger than the lateral ones, as is characteristic of other members of the genus. Adapted to the clay soils of bottomlands, this species is fast-growing, long-lived, and characteristic of the late stages of plant succession. Its distribution closely coincides with the drainage system of the Mississippi River and its major tributaries. It is widely planted as an ornamental and in bottomland plantations. Flowering is in April and May, with nut maturation and dehiscence of the husks in October. One of the most economically important trees of North America, *C. illinoinensis* is prized for both its nuts and its wood, the latter used for furniture, veneer paneling, and flooring. It is also prized as a shade tree. Horticulturalists have developed numerous cultivars. Comanches and Kiowas used

Leaves and nuts of *Carya illinoinensis*. Photo courtesy of the Samuel Roberts Noble Foundation.

the leaves and bark to treat ringworm and tuberculosis. The specific epithet has long been spelled *illinoensis* and appears this way in many taxonomic manuals. However, the original spelling by Friedrich Wangenheim, who described the species in 1787, is *illinoinensis*, and this is the spelling that must be used. In addition, the binomials *Hicoria pecan* and *Carya pecan* may be encountered in the older literature.

## *Castanea ozarkensis* Ashe

OZARK CHINKAPIN

**black, brown, red, yellow**

A member of the Fagaceae (beech family), *Castanea ozarkensis* is now encountered as a monoecious shrub up to 10 m tall that resprouts from its roots after dieback caused by chestnut blight. Before the introduction of the fungus *Cryphonectria parasitica* (formerly *Endothia parasitica*) from Asia in 1904 and the subsequent decimation of the species by the 1930s, large trees up to 20 m tall were typical. The fissured bark of this species is brownish, and the alternate, simple leaves are oblanceolate to obovate. The blade margins are serrate, with each tooth terminating in an awn. The pistillate flowers are borne in spiny involucres that completely enclose the nuts until mature. A deciduous forest species, *C. ozarkensis* occurs in the Ozark and Ouachita mountains of Missouri, Arkansas, Louisiana, Oklahoma, and Texas. Flowering is in June.

Leaves of *Castanea ozarkensis*.

## *Castilleja indivisa* Engelm.

INDIAN PAINTBRUSH, TEXAS PAINTBRUSH, ENTIRE-LEAF INDIAN PAINTBRUSH

**brown, green, red, yellow**

A member of the Scrophulariaceae (figwort family), *Castilleja indivisa* is an annual, semiparasitic herb that grows 20–40 cm tall from a slender taproot. Its erect stems are unbranched and bear alternate, simple leaves with linear to lanceolate blades. It is readily recognized by its showy inflorescence consisting of brightly colored bracts and sepals of various shades of red (rarely white or yellow). The corollas of five fused pale green-yellow petals are bilabiate. Its many-seeded capsules are ovoid. Characteristic of the

early stages of plant succession, *C. indivisa* is a prolific seed producer and generally establishes small to large populations in disturbed sites such as rights-of-way and heavily grazed pastures. Although photosynthetic, it is semiparasitic on the roots of associated grasses. Flowering occurs from late April to July. It occurs only in Oklahoma and Texas and is prized as a roadside wildflower, although its presence indicates disturbance, heavy grazing, or shallow soils. The common name "Indian blanket" has also been used for this species but is more properly applied to *Gaillardia pulchella*, the state wildflower of Oklahoma.

Inflorescence of *Castilleja indivisa*.
Photo by Wayne J. Elisens.

***Castilleja purpurea*** (Nutt.) G. Don **var. *citrina*** (Pennell) Shinners
YELLOW PAINTBRUSH, LEMON PAINTBRUSH, CITRON PAINTBRUSH, YELLOW INDIAN PAINTBRUSH
**brown, yellow**

A member of the Scrophulariaceae (figwort family), *Castilleja purpurea* var. *citrina* is a perennial, semiparasitic herb that grows 20–40 cm tall from a

Flowering plant of *Castilleja purpurea* var. *citrina*.
Photo by Robert J. O'Kennon, courtesy of the Botanical Research Institute of Texas.

woody rootstock. Its clustered, ascending to erect stems bear numerous alternate, simple leaves with linear to lanceolate, lobed blades. As its common names imply, this species is readily recognized by its showy inflorescence consisting of bright yellow or greenish yellow bracts and sepals. The lower lip of the bilaterally symmetrical corolla is 3–7 mm long and flaring. This species occurs as scattered plants or small populations in rocky or sandy soils of prairie hillsides from south-central Kansas through western Oklahoma to central Texas. Flowering is in April and May.

### *Catalpa speciosa* (Warder) Warder ex Engelm.

NORTHERN CATALPA, CIGAR-TREE, CATAWBA-TREE, HARDY CATALPA

**black, brown, yellow**

A member of the Bignoniaceae (catalpa family), *Catalpa speciosa* is a deciduous tree up to 30 m tall, with a pyramidal crown and red-brown bark broken into thick scales. Its opposite, simple leaves are long-petioled with large, broadly ovate to ovate-oblong, acuminate blades. Borne in terminal panicles, the showy flowers have bilabiate, whitish corollas with two yellow stripes and purple-brown spots in their throats. The pendulous, elongate, cylindrical, thick-walled capsules, 20–45 cm long, are diagnostic. Occurring naturally in moist woods and along streams of the Mississippi River drainage and its major tributaries, *C. speciosa* has been widely planted as an ornamental throughout the eastern half of the continent. It readily escapes cultivation and naturalizes. Flowering is in May and June.

Leaves and flowers of *Catalpa speciosa*.
Photo by Andrew Crosthwaite.

### *Ceanothus americanus* L.

NEW JERSEY TEA, REDROOT

**black, brown, green, orange, purple, red, yellow**

A member of the Rhamnaceae (buckthorn family), *Ceanothus americanus* is readily recognized by its shrubby habit from a massive rootstock; its alternate, simple, elliptic, three-nerved leaves; and its dense inflorescences of small white flowers borne on elongate peduncles. Recognition in winter or in vegetative condition is facilitated by the presence of persistent triangular to round disks at the ends of the pedicels. Small populations typically

Flowering plants of *Ceanothus americanus*.

occur in rocky or sandy soils of dry upland sites in open woods, glades, or prairie hillsides, and occasionally occur along roadsides. This species is an indicator of late stages of plant succession. Adapted to moderate, periodic disturbance such as fire and grazing, it declines with increasing disturbance and overuse by cattle. Plants are capable of nitrogen fixation. Distribution is throughout the eastern half of the continent as far west as Minnesota, eastern Nebraska, Kansas, Oklahoma, and Texas. Flowering is in May and June, with the capsules maturing in July and August. *Ceanothus americanus* is occasionally grown as an ornamental for its clusters of white flowers, which can be quite showy and fragrant. Native Americans and settlers used the dried leaves as a substitute for tea, although caffeine is not present in the leaves. The rootstocks were used as an astringent, an expectorant, and a wash for venereal sores and skin cancers. The bark of the rootstocks contains compounds with blood-clotting properties. Settlers referred to the massive rootstocks as "grubs" because they often had to be dug by hand before plowing could be completed.

### *Centaurea americana* Nutt.

BASKET FLOWER, AMERICAN STAR-THISTLE

**green, yellow**

A member of the Asteraceae (sunflower family), *Centaurea americana* is an annual herb that arises from stout taproots. Its erect stems are branched above, 30–150 cm tall, and bear alternate, simple leaves with lanceolate to ovate-lanceolate, entire or occasionally denticulate blades. The flowers are borne in large showy heads at the ends of upper branches. Only tubular disk florets are present, and the conspicuous articulated phyllaries give the heads the appearance of being woven baskets, hence the name "basket flower." Characteristic of the mid stages of plant succession and occurring in either small or large populations, *C. americana* is encountered in grasslands and roadsides in a variety of soil types. Often the soil is shallow and rocky. It is distributed in the south-central part of the United States and adjacent Mexico, with populations found from Missouri to Louisiana in the east, to Arizona in the west, and to Nuevo León and Coahuila in the south. Flowering occurs from late May to July. This plant is occasionally cultivated as an ornamental. Kiowas made poultices of the leaves to treat skin sores.

Head of *Centaurea americana*. Photo by George M. Diggs Jr.

## *Cephalanthus occidentalis* L.

BUTTONBUSH, HONEY-BALLS, BUTTON WILLOW, SPANISH PINCUSHION

**black, brown, green, red, yellow**

A member of the Rubiaceae (madder family), *Cephalanthus occidentalis* is a shrub or rarely a small irregularly shaped tree 1–4 m tall. Its glossy green leaves are whorled or opposite, simple, and have entire, elliptic or lanceolate-oblong blades with acuminate apices. The small, white to cream, funnelform, four-merous flowers are borne in showy, globose heads at the ends of long, often branched peduncles. In winter the globose clusters of brown or brownish green fruits persist. This species is a classic indicator of wetlands and is found at the edges of ponds, lakes, marshes, creeks, and intermittent streams. It is characteristic of the early stages of plant succession. Capable of with-

Inflorescences of *Cephalanthus occidentalis*. Photo by Terrence G. Bidwell.

standing complete submergence for several weeks at a time, it also occurs at the edges of reservoirs with fluctuating water levels. It is distributed throughout the eastern half of the continent as far west as eastern Nebraska, Kansas, Oklahoma, and Texas. Flowering is from May to September, and the flowers provide nectar and pollen for bees, hummingbirds, and butterflies. This plant is sometimes used as an ornamental. Native Americans used the bark to treat a variety of ailments, including fevers, bronchitis, venereal diseases, dysentery, eye problems, headaches, toothaches, nausea, and rheumatism.

## *Cercis canadensis* L.

REDBUD

**black, brown, green, orange, red, yellow**

A member of the Caesalpiniaceae (caesalpinia family), *Cercis canadensis* is a small or medium tree, up to 8 m tall, with a short trunk and spreading branches. It is readily recognized by its dark bark with scaly plates, its alternate, simple, cordate to reniform leaves; and its umbellate clusters of flowers with their pink, papilionaceous corollas borne on knotty spurs. In the winter it can be identified by its persistent flat legumes and knotty zigzag branches. Occurring either as scattered trees or small localized populations, *C. canadensis* is an understory species in open woods or their edges. Characteristic of the early to mid stages of plant succession, it typically occurs in moist loamy or sandy soils of valleys and bottomlands, or occasionally on moist upper slopes. It is distributed throughout the eastern half of the United States and extends west to eastern Nebraska, Kansas, Oklahoma, Texas, and northeastern Mexico. This plant is among the first species to flower in spring (usually April), with the flowers appearing before the leaves. Prized for the beauty of its flowers and for its glossy foliage, *C. canadensis* is a popular ornamental and a good humus builder. Its edible flowers can be fried or added fresh to salads as a garnish, and its tender young legumes can be sautéed in butter. In older references the genus is positioned in the subfamily Caesalpinioideae of the broadly circumscribed family Fabaceae (Leguminosae), the pea family.

Flowers of *Cercis canadensis*.
Photo by Wayne J. Elisens.

## *Chrysopsis pilosa* Nutt.

SOFT GOLDEN-ASTER

**black, green, orange, yellow**

A member of the Asteraceae (sunflower family), *Chrysopsis pilosa* is a taprooted, annual herb 30–80 cm tall. As the common name implies, the stems and leaves are pilose with a puberulent inner layer. The alternate, simple leaves are oblong to oblanceolate with entire to dentate or incised margins. The flowers are borne in numerous hemispheric heads at the ends of branches. Both the ray and disk florets are yellow, with the former flat and spreading or ascending during the day but curled and erect at night. The spindle-shaped achenes are pilose and bear a double pappus comprising inner bristles and outer scales. Characteristic of loose sandy soils in dry open sites, this species is distributed primarily in the south-central portion of the United States from Missouri and Kansas south to Texas and Louisiana. Some taxonomists classify it as *Bradburia pilosa* or *Heterotheca pilosa*.

Heads of *Chrysopsis pilosa*.

## *Cicuta maculata* L.

WATER-HEMLOCK, SPOTTED WATER-HEMLOCK, CHILDREN'S BANE, DEATH-OF-MAN, POISON PARSLEY, BEAVER POISON

**black, brown, orange, yellow**

A member of the Apiaceae (carrot family), *Cicuta maculata* is a perennial herb that grows up to 2.5 m tall from fleshy tuberous roots. Chambered near the bases, its erect stems are sometimes purple-streaked. The chambers contain an oily yellowish liquid that turns brown upon exposure to air. The alternate one-, two-, or three-pinnately compound leaves have lanceolate, serrate leaflets. Borne in large, open, hemispheric umbels, the small white flowers are five-merous and produce oval to orbicular, conspicuously ribbed schizocarps. Occurring across the continent, this species is readily recognized by its typically solitary occurrence along waterways, its tall stature, and its large white inflorescence. *Cicuta maculata* is among the most potent neurotoxic plants in North America. It acts quickly on the central nervous system, causing violent seizures before death. Well aware of its effects, Na-

Flowers and fruits of *Cicuta maculata*. Photo by George M. Diggs Jr.

tive Americans used it ceremonially and medicinally to treat rheumatism and skin and orthopedic problems.

### *Cirsium undulatum* (Nutt.) Spreng.

WAVY-LEAF THISTLE

**black, brown, green, yellow**

A member of the Asteraceae (sunflower family), *Cirsium undulatum* is a perennial herb 30–100 cm tall. Its stems and alternate, simple leaves are conspicuously white-tomentose. Lanceolate to oblanceolate, the sessile blades are pinnatifid, with each lobe tipped by a stout yellow spine. The large heads are solitary at the ends of the branches and bear only pink-purple or whitish purple tubular disk florets. When mature, the achenes have a pappus of white plumose bristles. Occurring as scattered plants or small localized populations in upland prairies, pastures, roadsides, and other occasionally disturbed sites, this species is distributed in the western two-thirds

Heads and upper leaves of *Cirsium undulatum*. Photo by Robert J. O'Kennon, courtesy of the Botanical Research Institute of Texas.

of the United States. Native Americans ate the young plants and used them to make a general tonic and decoctions and infusions to treat eye and venereal diseases.

### *Comandra umbellata* (L.) Nutt.

BASTARD TOAD-FLAX, WESTERN COMANDRA

**brown, orange, yellow**

A member of the Santalaceae (sandalwood family), *Comandra umbellata* is an erect, semiparasitic, perennial herb or subshrub 7–50 cm tall that arises from rhizomes. Typically gray-green, its alternate, simple leaves are linear to elliptic or lanceolate and subsessile to short-petiolate. Borne in terminal cymes, the flowers are radially symmetrical, have a uniseriate perianth, and are white with a green tube. They produce a drupe-like one-seeded fruit. Although photosynthetic, this species parasitizes via haustoria the roots of nearby species. It occurs across the continent, typically in dry rocky or sandy soils in prairies and open upland woods. Native Americans used it to make infusions and decoctions to treat eye problems, canker sores, breathing problems, and headaches.

Flowers of *Comandra umbellata*.
Photo by Wayne J. Elisens.

### *Coreopsis tinctoria* Nutt.

PLAINS TICKSEED, PLAINS COREOPSIS, GOLDEN TICKSEED

**black, brown, green, orange, yellow**

A member of the Asteraceae (sunflower family), *Coreopsis tinctoria* is a glabrous, annual herb 40–120 cm tall. Its erect stems are branched above, with the branches stiffly ascending. The alternate, one- or two-pinnately divided or dissected, entire leaves have linear lobes. Flowers are borne in showy heads at the ends of branchlets. The seven or eight three-lobed ray florets are distally yellow and basally red or reddish brown, whereas the numerous disk florets are yellow. In fruit, the flattened achenes are winged or not winged. A pappus of two minute awns may or may not be present. Characteristic of the early stages of plant succession, this species occupies a variety of soil types, primarily clays and loams. Its habitats are generally wet or damp and include borrow ditches, low-lying fields, and pond margins. It thrives on disturbance and may multiply following fire. A prolific achene producer, it typically forms large populations and often dominates a site. *Coreopsis tinctoria* is now encountered across the continent, although its original distribution is believed to be the central grasslands. It is frequently cultivated, escapes, and thus occurs sporadically elsewhere. Flowering is from July to October. This species is cultivated as an ornamental. It provides cover and food for bobwhite quail and is occasionally browsed by white-tailed deer, though it is of low preference. Systematic work reveals that *C. tinctoria* is conspecific with *C. cardaminaefolia*, the two taxa differing morphologically only in one Mendelian character.

Heads of *Coreopsis tinctoria*. Photo by George M. Diggs Jr., courtesy of the Botanical Research Institute of Texas.

### ***Cornus drummondii*** C.A. Mey.

ROUGH-LEAF DOGWOOD, ROUGH-LEAVED DOGWOOD

**black, brown, green, orange, yellow**

A member of the Cornaceae (dogwood family), *Cornus drummondii* is a shrub 1–2 m tall that typically forms thickets from root sprouts. It is readily recognized by its reddish brown twigs, its opposite, rough-surfaced leaves, and its flat-topped cymes of cream-white flowers and white drupes. Also diagnostic are the cottony strands of vascular tissue that unite halves of the leaf blades when they are gently torn apart transversely. In winter it can be identified by its reddish brown twigs and by the remnants of its cymes, bearing an occasional drupe. Unlike *C. florida* (flowering dogwood), this species lacks showy bracts. It is a classic indicator of openings in woods and the interface between grassland and forest. Characteristic of the early

Flowers of *Cornus drummondii*.

stages of plant succession, it thrives with occasional site disturbance. This species is distributed throughout the Midwest from the Ohio and Mississippi river valleys to southeastern South Dakota, eastern Nebraska, Kansas, Oklahoma, and central Texas. Flowering is from late April to early June. The drupes mature and turn white in late summer. The thickets provide cover and the drupes food for many birds and mammals. The foliage and twigs are moderately preferred by white-tailed deer and may be especially important in areas where the animals are overpopulated. Native Americans used the plants to prepare remedies for diarrhea, toothaches, and muscle aches. The branches were used as arrow shafts, and the twigs served as chew sticks with which to massage gums and clean teeth. The binomial *C. asperifolia* was used for this species in older publications. *Cornus drummondii* is reported to hybridize with *C. amomum* (silky dogwood) and *C. foemina* (stiff dogwood).

## *Cornus florida* L.

FLOWERING DOGWOOD

**black, brown, green, orange, yellow**

A member of the Cornaceae (dogwood family), *Cornus florida* is a small, often somewhat crooked tree up to 12 m tall. Its bark is dark, closely checkered, and resembles the skin of an alligator. Borne on green branchlets, its opposite, simple, glossy green leaves have elliptic to ovate, entire blades with acuminate apices. Borne in dense clusters, the small greenish cream to white flowers are subtended by four large, showy, white bracts and produce clusters of red to orange ellipsoid drupes. Also diagnostic are the cottony

Flowers and subtending bracts of *Cornus florida*. Photo by Andrew Crosthwaite.

strands of vascular tissue that unite halves of the leaf blades when they are gently torn apart transversely. A fast-growing, short-lived understory climax species, *C. florida* is found in moist soils of the eastern deciduous forest throughout the eastern half of the continent. One of North America's contributions to the world of horticulture, it is widely planted as an ornamental. Its wood is very hard, strong, shock-resistant, and with continuous use becomes very smooth. At one time, shuttles for textile weaving, spools, small pulleys, and jeweler's blocks were made of this wood. Native Americans made infusions and decoctions of the bark to treat headaches, fevers, worms, measles, and other problems and diseases. In the 1800s the root bark was used in place of quinine to treat malaria. Birds eat the drupes and deer browse the foliage.

## *Cotinus obovatus* Raf.

SMOKE-TREE, AMERICAN SMOKE-TREE, SMOKEWOOD

**black, brown, orange, yellow**

A member of the Anacardiaceae (cashew family), *Cotinus obovatus* is a shrub or small deciduous tree up to 10 m tall, with yellow wood, scaly bark, and strong-smelling sap. Its brown branches bear alternate, simple, obovate to elliptic leaves that turn bright yellow to orange, red, or scarlet in autumn. The numerous small, yellowish white to greenish yellow, imperfect flowers are borne in terminal panicles intermixed with sterile, plumose pedicels that give the inflorescences a wispy smoke-like appearance, hence the common names. Although this plant has many flowers, only a few reniform drupes are produced per inflorescence. Occurring as solitary or scattered plants in rocky woods or limestone outcrops, this species is distributed in the southeastern quarter of the United States as far west as eastern Oklahoma and central Texas. *Cotinus coggygria*, its Old World relative, is widely cultivated as an ornamental.

Inflorescence of *Cotinus obovatus*.
Photo by Bruce W. Hoagland.

## ***Cucurbita foetidissima*** Kunth

BUFFALO GOURD, STINKING GOURD, FOETID GOURD

**brown, green, yellow**

A member of the Cucurbitaceae (gourd family), *Cucurbita foetidissima* is a perennial vine with multiple prostrate or trailing stems extending 6 m or more from a large spindle-shaped root. The large, thick, deltoid to ovate, irregularly toothed, gray-green leaves are foul-smelling when crushed. Borne singly in leaf axils, the large yellow flowers produce globose pepos up to 7.5 cm in diameter that are at first green with lighter stripes but lemon-yellow at maturity. Occurring in sandy and rocky soils of disturbed sites, especially roadsides and fencerows, this species is distributed from Missouri and Nebraska south to Texas, northern Mexico, Arizona, and southern California. Plains tribes and western tribes used *C. foetidissima* in a variety of ways—as medicines, ceremonial rattles, toys, and cleansers.

## ***Cycloloma atriplicifolium*** (Spreng.) Coult.

TUMBLE RINGWING, WINGED PIGWEED

**green, yellow**

A member of the Chenopodiaceae (goosefoot family), *Cycloloma atriplicifolium* is an annual herb, typically with multiple arching, branched stems that give the plant a globose, bushy, tumbleweed-like appearance. Its alternate, simple, sessile or short-petioled leaves have lanceolate to ovate, irregularly sinuate-dentate or pinnately lobed blades. Borne widely spaced in paniculate spikes, the small flowers have five sepals, each with a broad, horizontal, white-hyaline, irregularly lobed wing. This ring of sepal wings persists around the mature, black, shiny utricle and becomes red-purple with age. Typically encountered in loose sandy soils of dunes, roadsides, wheat fields, and other disturbed sites, this species occurs in the Midwest from Manitoba and Indiana south to Texas and Arizona. It is adventive in the eastern United States and Europe. Native Americans used this plant as food, as a treatment for headaches and rheumatism, and in ceremonies.

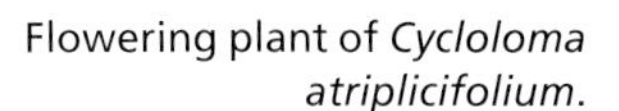

Flowering plant of *Cycloloma atriplicifolium*.

### *Cyperus squarrosus* L.

BEARDED FLATSEDGE

**brown, green, orange, yellow**

A member of the Cyperaceae (sedge family), *Cyperus squarrosus* is a sweet-scented, cespitose, annual herb 2–13 cm tall. Its few leaves are basal and subtend the inflorescence, which consists of a sessile or umbellate cluster of spikes subtended by long bracts. Conspicuously flattened, the ovate to oblong, greenish to reddish brown spikelets bear eight to sixteen two-ranked flowers in the axils of seven- or nine-nerved scales with recurved apices. The trigonous, light brown achenes terminate in a three-cleft style. Typically forming large populations in moist, disturbed soils at the edges of streams, ponds, and temporary puddles, this species is distributed across the continent.

### *Dalea candida* Michx. ex Willd.

WHITE PRAIRIE-CLOVER

**brown, green, yellow**

A member of the Fabaceae (pea family), *Dalea candida* is a perennial herb 30–100 cm tall that arises from a woody rootstock and bears alternate, one-pinnately compound leaves with three or five linear-oblong leaflets. Borne in terminal, ovoid to cylindrical spikes about 1 cm in diameter, the white flowers occur in the axils of small bracts. Unlike other members of the family, its corollas are not papilionaceous but rather radially symmetrical or slightly bilateral with one petal somewhat larger than the others. The one- or two-seeded legumes are beaked and villous. Characteristic of the late stages of plant succession, *D. candida* is found on a variety of soils in prairies and in open, rocky, upland woods. It is distributed primarily in the center of the continent from Canada to northern Mexico. Flowering is from May through September. Native Americans used this plant in the treatment of abdominal pain and toothaches. They also ate the roots and incorporated the plant in various religious ceremonies. Some taxonomists

Flowering and fruiting inflorescences of *Dalea candida*. Photo courtesy of the Samuel Roberts Noble Foundation.

classify this species as *Petalostemon candidus* because it differs from other species of *Dalea* in stamen number and corolla symmetry.

## *Dalea lanata* Spreng.

WOOLLY DALEA, WOOLLY PRAIRIE-CLOVER

**brown, green, yellow**

A member of the Fabaceae (pea family), *Dalea lanata* is a densely villous-to-mentose, perennial herb that arises from long, orange, woody roots. Its prostrate to ascending-trailing stems are 30–50 cm long and bear alternate, one-pinnately compound leaves with nine to thirteen obovate, emarginate leaflets. Borne in peduncled spikes arising in the leaf axils, the papilionaceous corollas are red to purple. Its one-seeded legumes tend to remain wholly or partially enclosed in the villous calyces. Occurring as solitary plants or small populations, this species is characteristically found in sandy soils of dunes and river bottoms. Its general distribution is from Utah, Colorado, and Kansas south to Arizona, New Mexico, and Texas, with some populations in Nevada and Arkansas. Flowering is from June to October. Native Americans used it in poultices to treat insect bites and ate the roots as candy.

Flowering plants of *Dalea lanata*.
Photo by Bruce W. Hoagland.

## *Dalea purpurea* Vent.

PURPLE PRAIRIE-CLOVER

**green, yellow**

A member of the Fabaceae (pea family), *Dalea purpurea* is a perennial herb 30–100 cm tall, with many erect or ascending stems arising from woody rootstocks. Its herbage has a citrus-like odor, and the alternate, one-pinnately compound leaves are conspicuously black and gland-dotted, bearing five or rarely three or seven linear leaflets. Borne in terminal, oblong spikes 1–1.5 cm in diameter, the purple or rose-purple flowers occur in axils of brown, silky-villous bracts. Unlike those of other members of the family, the corollas of this species are not papilionaceous but radially symmetrical or slightly bilateral with one petal somewhat larger. The one- or two-seeded legumes are beaked and villous. Characteristic of the late stages of plant succession, *D. purpurea* occurs as either scattered plants or in small populations on a variety of soils in prairies and open, rocky, upland woods. It is sometimes associated with limestone sites. It increases in abundance and individual plant vigor after winter fires. It is distributed primarily in the Midwest from Canada south to Texas and New Mexico and from Tennessee and Arkansas west to Colorado and Montana. Populations also occur as far east as New York. Flowering is from June to September. Native Americans used it in teas and medicinal decoctions and chewed the roots as a candy. The foliage is high in protein, especially when young, and is consumed by both livestock and wildlife. Some taxonomists classify this species as *Petalostemon purpureus* because it differs from other species of *Dalea* in stamen number and corolla symmetry. The binomial *Petalostemum purpureum* also appears in the literature; the difference in spelling of both the generic name and the specific epithet is due to different opinions of the etymology of the generic name.

Inflorescences of *Dalea purpurea*.
Photo by Adam K. Ryburn.

## *Dalea villosa* (Nutt.) Spreng.

SILKY PRAIRIE-CLOVER

**brown, green, yellow**

A member of the Fabaceae (pea family), *Dalea villosa* is a densely villous, perennial herb 20–40 cm tall that arises from branching, red-orange rootstocks. Its ascending to erect, solitary or multiple stems bear alternate, one-pinnately compound leaves with eleven to twenty-one elliptic to oblan- ceolate leaflets. Borne

Inflorescences of *Dalea villosa*.
Photo by Bruce W. Hoagland.

in terminal oblong-cylindrical spikes 7–10 mm in diameter, the rose-purple or pink to nearly white flowers occur in the axils of silky-villous bracts. Unlike those of other members of the family, the corollas are not papilionaceous but radially symmetrical or slightly bilateral with one petal somewhat larger. The one-seeded legumes are villous. Characteristically found in loose sandy soils of dunes, river valleys, and sandy prairies, this species is distributed primarily in the Rocky Mountains, Great Plains, and tallgrass prairie from Canada south to Texas and New Mexico. Flowering is from June to August. Native Americans used the roots as a purgative and the leaves to treat throat problems. Some taxonomists classify this species as *Petalostemon villosus* because it differs from other species of *Dalea* in stamen number and corolla symmetry.

### ***Desmanthus illinoensis*** (Michx.) MacMill. ex B.L. Rob. & Fernald

ILLINOIS BUNDLEFLOWER, PRAIRIE MIMOSA, PRAIRIE BUNDLEFLOWER

**black, brown, green, yellow**

A member of the Fabaceae (pea family), *Desmanthus illinoensis* is a perennial herb 30–120 cm tall that arises from woody rootstocks. It is readily recognized by its two-pinnately compound leaves bearing numerous small leaflets that fold upon contact, and by its globose spikes of white flowers with their conspicuous exserted stamens. It can also be identified by the dark,

Flowering plants of *Desmanthus illinoensis*. Photo by Charles S. Lewallen.

globular clusters of strongly curved legumes that persist after the leaves have dropped. Present in all stages of plant succession, this species typically occurs in a variety of wet or dry soils in prairies and open woods. Large populations often appear in borrow ditches and other disturbed sites. Following frequent fires, it may increase in abundance. It is distributed primarily in the central United States from Alabama and New Mexico north to Ohio, Minnesota, and the Dakotas. Flowering occurs from June to August. Native Americans used the clusters of legumes for children's rattles and boiled the leaves to remedy itches. Cattle, white-tailed deer, and pronghorn antelope relish this plant, which decreases with heavy grazing. It is high in protein and produces abundant seeds that are consumed by bobwhite quail, prairie chickens, wild turkeys, numerous songbirds, and small mammals. Considered to be among the most important native legumes, it is used in rangeland revegetation programs.

### ***Desmodium glutinosum*** (Muhl. ex Willd.) A.W. Wood

STICKY TICKCLOVER, LARGE-FLOWERED TICKCLOVER, POINTED-LEAF TICKCLOVER

**brown, yellow**

A member of the Fabaceae (pea family), *Desmodium glutinosum* is an erect, perennial herb 40–100 cm tall. Its stems are unbranched below the inflorescence and bear alternate, one-pinnately compound, subsessile leaves that have three oblong to lanceolate leaflets. Borne in terminal, elongate racemes, the showy flowers have papilionaceous corollas with pinkish lavender or white petals. As is characteristic of the genus, the fruits are flattened, curved loments that are conspicuously constricted between the seeds and break into one-seeded, triangular, densely uncinate segments when mature. Occurring either as scattered plants or in small localized populations, this species is characteristic of the mid to late stages of plant succession and is found in dry, rocky, open woods. Its distribution is the eastern

Flowering plant of *Desmodium glutinosum*. Photo by Dan Tenaglia.

half of the continent. Flowering is from June to August, and the seeds are dispersed via the hooked hairs of the loment segments, which readily cling to clothing, skin, or fur. Native Americans used the roots and leaves in medicinal infusions. The seeds are high in protein and calcium and are eaten by bobwhite quail, wild turkeys, ring-necked pheasants, and white-footed mice. Bees visit the flowers for nectar.

## *Diospyros virginiana* L.

PERSIMMON

**black, brown, orange, yellow**

A member of the Ebenaceae (ebony family), *Diospyros virginiana* is a small to large, dioecious or rarely polygamous-dioecious tree up to 20 m tall that occurs singly or in thickets. Its bark is dark brown-black and deeply furrowed into square plates. Lacking terminal buds, the twigs are slender, zigzag, and bear alternate, simple, more or less conduplicate leaves with elliptic or ovate, coriaceous blades that are dark green and glossy above and whitish green below. Its inflorescences are of two types: staminate, with clusters of two- to three-flowered cymes; and pistillate, with solitary flowers on short recurved pedicels and subtended by bracts. The imperfect flowers are fourmerous, and the pistillate ones produce large, subglobose, yellow or pale orange, glaucous berries with eight oblong, thick, flat, hard, brown seeds. Slow-growing and short-lived, this species occupies a variety of soils and habitats, including rocky dry uplands and swampy bottomlands, and is characteristic of the early to mid stages of plant succession. It is capable of infesting overgrazed, introduced pastures, old fields, and abandoned croplands. It sprouts from its roots, especially after fire, and often forms dense thickets of tall, straight, spindly trees. Flowering is in May and June. This species is distributed in the eastern half of the United States from New York south to Florida and west to eastern Kansas, Oklahoma, and Texas. The fruits are astringent when green but sweet and edible when

Leaves and fruits of *Diospyros virginiana*.
Photo by Steven K. Goldsmith.

ripe (usually after killing frosts in fall) and were used by Native Americans and early settlers for food and medicine. The pulp was smashed, dried, and used as an additive to many other foods. Dried leaves were used to make teas. The wood is used for golf-club heads, tool handles, and furniture veneer. The fruits are also important food for wildlife.

### ***Dracopis amplexicaulis*** (Vahl) Cass.

CLASPING-LEAVED CONEFLOWER, CLASPING CONEFLOWER, CLASPING BROWN-EYED SUSAN

**black, brown, green, red, yellow**

A member of the Asteraceae (sunflower family), *Dracopis amplexicaulis* is an erect, glabrous, annual herb 30–70 cm tall. Its alternate, simple, oblong to ovate leaves are sessile, with their bases clasping the stem. Flowers are borne in large, showy, hemispheric heads at the ends of branches. The five to nine ray florets are yellow or yellow and red-brown to brown-purple at their bases, whereas the numerous disk florets are purplish brown and subtended by stiff receptacular bracts. When mature, the achenes are terete and lack a pappus. Typically forming large populations in moist, disturbed sites such as roadside ditches and drainage ways, this species occurs primarily in states of the Gulf Coast, from Georgia and Florida to Texas and Oklahoma, but can also be found as far north as Missouri and as far west as New Mexico. Flowering is from May to July, and its achenes are eaten by a number of songbirds. Use by Native Americans is not reported. Some taxonomists classify this species as *Rudbeckia amplexicaulis*.

Heads of *Dracopis amplexicaulis*. Photo by Robert J. O'Kennon, courtesy of the Botanical Research Institute of Texas

### ***Echinacea angustifolia*** DC.

PURPLE PRAIRIE CONEFLOWER, BLACKSAMPSON, ECHINACEA, PRAIRIE CONEFLOWER

**black, brown, green, yellow**

A member of the Asteraceae (sunflower family), *Echinacea angustifolia* is an erect, coarsely hirsute, perennial herb 10–50 cm tall that arises from vertical woody rootstocks. Its alternate, simple leaves have narrowly elliptic-lan-

Population of *Echinacea angustifolia*.

ceolate to oblong blades, with the basal ones short- to long-petioled and the upper ones subsessile. The flowers are produced in large, showy, solitary heads borne at the ends of elongate peduncles. The numerous drooping, two- or three-lobed ray florets are light purple, magenta, or pink, whereas the numerous disk florets are yellow to purple and subtended by stout receptacular bracts. Bearing a short-toothed, crown-like pappus, the mature achenes are four-sided. This species is typically encountered in loamy and clay-loamy soils and is especially abundant in the mid to late stages of plant succession in the tallgrass prairie. Its abundance declines when cattle grazing is continuous. This plant also thrives in areas with occasional disturbance, such as in the margins of highway rights-of-way. Flowering is from June to August. *Echinacea angustifolia* is distributed throughout the tallgrass prairie and Great Plains from Saskatchewan and Minnesota south to Texas. Native Americans and early European colonists used this species medicinally to treat a plethora of problems. Today it is incorporated in a variety of medicines and herbal remedies. Unfortunately, the rootstocks of wild plants of *E. angustifolia* are worth considerable money. For this reason they are harvested and sold, and consequently many populations throughout the species' range have been extirpated. These plants are also cultivated as garden ornamentals. Some taxonomists treat the species as a variety of *E. pallida*. The common name "coneflower" is also applied to species of *Ratibida*, a member of the same family.

### ***Echinocereus reichenbachii*** (Terscheck ex Walp.) F. Haage

LACE HEDGEHOG-CACTUS, LACE CACTUS, WHITE LACE CACTUS

**green, yellow**

A member of the Cactaceae (cactus family), *Echinocereus reichenbachii* is a small barrel-shaped cactus 7–15 cm tall and 3–6 cm in diameter. Its stems are ribbed and bear numerous areoles with twelve to thirty-two spreading radial spines that hide the surface and impart a lacy appearance, hence the common names. The large showy flowers are 5–8 cm in diameter with numerous pink to purple perianth parts and produce globose to ovoid green berries with woolly areoles. Encountered as solitary plants or small populations in gravelly, rocky, or sandy soils of limestone or granite outcrops, this species is distributed from southern Colorado and Oklahoma south to Texas and Mexico. Although its specific use by Native Americans has not been reported, other species of the genus were eaten, in particular the fruits. This species is named for Heinrich Gottlieb Reichenbach, an early-nineteenth-century German naturalist.

Flowers of *Echinocereus reichenbachii*.
Photo by Robert J. O'Kennon, courtesy of the Botanical Research Institute of Texas.

### ***Elephantopus carolinianus*** Willd.

LEAFY ELEPHANT'S FOOT, CAROLINA ELEPHANT'S FOOT

**black, brown, green, yellow**

A member of the Asteraceae (sunflower family), *Elephantopus carolinianus* is a coarsely hirsute-strigose, perennial herb that arises from fibrous rootstocks. The erect stems are 40–100 cm tall, branched above, and bear alternate, simple leaves with dark green, elliptic to oblanceolate

Heads and subtending bracts of *Elephantopus carolinianus*.

blades with attenuate bases. The leaves often form a basal rosette. Borne in glomerules of small cylindrical heads subtended by three deltoid bracts, the three or four flowers are unequally five-cleft and white, pinkish white, or violet. The cylindrical, ribbed achenes have a pappus of five to ten short rigid bristles with narrowly triangular bases. Characteristic of the mid to late stages of plant succession, *E. carolinianus* is a classic indicator of shaded stream terraces and bottomlands of eastern closed-canopy forests. Plants are typically found in moist loam or loamy-sand soils, often forming large populations. This species is distributed in the eastern half of the United States as far north as New Jersey, Pennsylvania, and Ohio and as far west as southeastern Kansas and eastern Oklahoma and Texas. Flowering is in August and September. Use by Native Americans has not been reported. Bobwhite quail and wild turkeys eat the achenes, and deer occasionally consume the heads and foliage, but this plant is a low-preference browse.

## *Erigeron strigosus* Muhl. ex Willd.

DAISY FLEABANE, WHITETOP, ROUGH FLEABANE, PRAIRIE FLEABANE

**brown, yellow**

A member of the Asteraceae (sunflower family), *Erigeron strigosus* is an annual or rarely biennial, strigose or hirsute herb that grows 30–70 cm tall from a fibrous root system. It is readily recognized by its erect stems that are unbranched below and branched above; its subsessile, cauline, oblanceolate to linear leaves; and its numerous heads with white ray florets and yellow disk florets. Characteristic of the early to mid stages of plant succession, this species occupies a variety of dry soils in pastures and fields. It is frequently encountered in disturbed sites and waste areas and occurs across the United States and southern Canada, with greatest abundance in the eastern half. Flowering is from May to late June. Native Americans used the leaves of *E.*

Heads of *Erigeron strigosus*.
Photo by Terrence G. Bidwell.

*strigosus* as a snuff to clear head colds and in infusions to treat headaches, heart problems, and mouth sores. As the common names imply, its smoke was reputed to be useful in getting rid of fleas and gnats. This plant increases in abundance with heavy grazing. It is not eaten by cattle but is a low- to moderate-preference browse for white-tailed deer. When abundant, it provides limited cover for bobwhite quail and small mammals.

### *Euphorbia maculata* L.

SPOTTED SPURGE, MILK SPURGE, PROSTRATE SPURGE, MILK PURSLANE

**black, green, yellow**

A member of the Euphorbiaceae (spurge family), *Euphorbia maculata* is a prostrate, annual, taprooted herb that produces a milky sap. Its multiple radiating stems are coarsely villous, 10–50 cm long, and bear opposite, simple, oblong leaves, each typically with a reddish or purple spot. The small imperfect flowers are borne in solitary cyathia in the leaf axils. These cyathia have four glands and white or reddish appendages. When mature, the capsular schizocarp is conspicuously three-lobed. Typically occurring in disturbed, sandy soils in woods and fields, along roadsides, and in waste areas, this species is believed to be native to the eastern half of the continent but is now naturalized as a weed elsewhere in the world. In sensitive individuals the milky sap of *E. maculata* may cause a dermatitis akin to photosensitization. The binomial *E. supina* is a synonym that frequently appears in the older literature. In addition, some taxonomists now classify this species as *Chamaesyce maculata*.

Leaves and cyathia of *Euphorbia maculata*. Photo courtesy of the Samuel Roberts Noble Foundation.

### *Euphorbia marginata* Pursh

SNOW-ON-THE-MOUNTAIN

**brown, green, yellow**

A member of the Euphorbiaceae (spurge family), *Euphorbia marginata* is an erect, taprooted, annual herb 30–100 cm tall that produces a milky sap. Its alternate, simple leaves are sessile with oblong, ovate, or elliptical blades. Small and imperfect, the flowers are produced in numerous cyathia borne in terminal cymes. The campanulate, villous cyathia are subtended by conspicuous, white-margined bracts and have five glands with large white petaloid appendages. When mature, the capsular schizocarp is conspicuously three-lobed. This species is a dominant of the early stages of plant succession. Typically occurring on clayey soils, it is a classic indicator of recent disturbance, often found at construction sites, in waste areas, and in overgrazed pastures. It occurs in the central United States from Montana and Minnesota south to Texas and New Mexico. Flowering is from July to October. Contact with the milky sap causes dermatitis in sensitive individuals. Native Americans used the plants as a liniment for swellings and as a tea to increase milk production in nursing mothers. It is cultivated as a garden ornamental and is a good wildlife species. The seeds, with their high lipid content, are eaten by a variety of birds. Although the foliage is eaten by pronghorn antelope, it is toxic to livestock and normally not eaten. The morphologically similar *E. bicolor* (snow-on-the-prairie) is distinguished by its narrower bracts. When the two species occasionally occur together, putative hybrids are sometimes encountered.

## ***Eustoma exaltatum*** (L.) G. Don

PRAIRIE GENTIAN, CATCHFLY GENTIAN, BLUEBELL GENTIAN, PURPLE PRAIRIE GENTIAN, TEXAS BLUEBELLS

**brown, yellow**

A member of the Gentianaceae (gentian family), *Eustoma exaltatum* is an erect, annual or short-lived perennial herb 25–75 cm tall that arises from a taproot. The opposite, simple, glaucous leaves have ovate to lanceolate or elliptic, entire blades with three prominent nerves. Borne in terminal, paniculate cymes, the conspicuously showy, five-merous flowers have blue-purple, pink, yellow, or white campanulate corollas. At maturity the ellipsoidal capsule is about 2 cm long. Occurring as solitary plants or small populations in moist prairie sites, this species is distributed in the tallgrass prairie and Great Plains from South Dakota and Nebraska south to Texas and northern Mexico. Flowering is from June to Octo-

Flowers of *Eustoma exaltatum*.
Photo by George M. Diggs Jr.

ber. Taxonomists differ in their opinions as to the number of species in the genus. Some recognize two species and use the binomials *E. exaltatum* and *E. grandiflorum* or *E. russellianum*.

### *Fraxinus americana* L.

WHITE ASH

**brown, yellow**

A member of the Oleaceae (olive family), *Fraxinus americana* is a dioecious medium tree up to 20 m tall. Its mature gray-brown to dark gray bark is deeply longitudinally furrowed, the furrows intersecting to form Ys. The opposite, one-pinnately compound leaves have five, seven, or nine oblong to elliptic leaflets that are glossy green above and white or pale green-white below. The leaf scars are broadly U-shaped and partially surround the axillary buds. The inconspicuous, imperfect flowers are borne in dense fascicles or panicles. When mature, the pendulous clusters of elongate samaras facilitate recognition at a distance. The wings of the samaras are terminal or decurrent less than a third the length of the samara body. Characteristic of moist soils (lower slopes, stream banks, floodplains) of the eastern deciduous forest, this species is distributed throughout the eastern half of the United States and reaches its westernmost limits in eastern Texas, Oklahoma, Kansas, and Nebraska. Flowering is from March to May, the samaras persisting on the trees until the following autumn and winter. The wood of this species is noted for its strength and shock resistance and is famous for use in baseball bats, hockey sticks, and other athletic equipment. Frames of upholstered furniture and veneer for cabinets are often made of this wood. Native Americans used this species extensively, making infusions and decoctions of the roots, bark, leaves, and flowers to treat a variety of ailments and using the branches and wood to make baskets, canoe frames, snowshoes, and implement handles. Like other species of the genus, *F. americana* is only moderately important to wildlife. The samaras are eaten by birds and small mammals.

### *Fraxinus pennsylvanica* Marsh

GREEN ASH, RED ASH

**brown, green, yellow**

A member of the Oleaceae (olive family), *Fraxinus pennsylvanica* is a dioecious medium tree up to 25 m tall. Its mature gray-brown to dark gray bark is deeply longitudinally furrowed, the furrows intersecting to form Ys. The opposite, one-pinnately compound leaves have five, seven, or nine oblong to elliptic leaflets that are glossy green above and green below. The leaf scars are broadly semicircular with the axillary buds borne above or in a small notch at the top. The inconspicuous, imperfect flowers are borne in dense fascicles or panicles. When mature, the pendulous clusters of elongate samaras facilitate recognition at a distance. The wings of the samaras are decurrent one-half to three-quarters the length of the samara body. Characteristic of moist soils (lower slopes, stream banks, floodplains) of the eastern deciduous forest, this species is also found in ravines in the tallgrass prairie. It is distributed throughout the eastern half of the United States and reaches its westernmost limits in the foothills of the Rocky Mountains. Flowering is from March to May, the samaras persisting on the trees until the following autumn and winter. The wood of this species is similar to that of *F. americana* and is used for athletic equipment, furniture frames, cabinet veneer, and firewood. This fast-growing species produces brilliant yellow leaves in autumn and is planted extensively as an ornamental. Native Americans used it in a variety of ways, such as to make medicines, bows, arrows, pipestems, implement handles, and ceremonial items. *Fraxinus pennsylvanica* is only of moderate importance to wildlife. The samaras are eaten by birds and small mammals.

Leaves and samaras of *Fraxinus pennsylvanica*.

## ***Fraxinus quadrangulata*** Michx.

BLUE ASH

**black, brown, green, yellow**

A member of the Oleaceae (olive family), *Fraxinus quadrangulata* is a large polygamous tree up to 30 m tall. Its mature tan to gray-brown bark is shallowly furrowed longitudinally, the furrows intersecting to form Ys. Borne on four-angled or four-winged twigs, the opposite, one-pinnately compound leaves have seven, nine, or eleven lanceolate to ovate, serrate leaflets. Its leaf scars are broadly crescent-shaped to semicircular with the axillary buds borne above or in a small notch at the top. The inconspicuous, imperfect flowers are borne in dense fascicles or panicles. When mature, the pendulous clusters of elongate samaras facilitate recognition at a distance.

The wings of the samaras are decurrent more than half the length of the samara body. Characteristic of moist soils of the eastern deciduous forest, this species is distributed from southern Ontario, Michigan, and Ohio south to northwestern Georgia and northern Alabama, and west to southeastern Kansas and northeastern Oklahoma. Flowering is in March and April. This species is an important timber tree, and its wood is used for athletic equipment, furniture frames, and cabinet veneer. Specific use by Native Americans is not reported. It is only of moderate importance to wildlife. The samaras are eaten by birds and small mammals.

Leaves of *Fraxinus quadrangulata*.
Photo by Bruce W. Hoagland.

## ***Froelichia floridana*** (Nutt.) Moq.

FIELD SNAKE-COTTON

**black, brown, yellow**

A member of the Amaranthaceae (pigweed family), *Froelichia floridana* is a white-canescent or white-tomentose, annual herb 40–100 cm tall. Its erect, somewhat flexuous stems bear opposite, simple leaves with oblong or elliptic to oblanceolate or spathulate blades. Borne in terminal, whitish, spicate inflorescences, the flask-shaped flowers are subtended by scarious bracts and bractlets and are essentially hidden in a mass of tangled hairs arising from their five fused sepals. The petals are absent and the filaments of the five stamens are fused to form a scarious tube. The fruits are utricles that remain enclosed in the calyces and staminal tubes.

Inflorescences of *Froelichia floridana*.

Characteristically encountered in loose sandy soils of dunes, sandy bottoms, and sandy prairies, this species is distributed primarily in the Midwest from South Dakota, Minnesota, Illinois, and Indiana south to Arkansas, Texas, and New Mexico. Flowering is from July to October. Use by Native Americans is not reported, nor is use by wildlife.

## *Gaillardia aestivalis* (Walter) H. Rock

PRAIRIE GAILLARDIA, LANCELEAF GAILLARDIA, YELLOW INDIAN BLANKET

**brown, green, yellow**

A member of the Asteraceae (sunflower family), *Gaillardia aestivalis* is an annual or short-lived perennial herb 30–70 cm tall. Branched above, its erect stems bear alternate, simple, mostly sessile leaves with oblong to lanceolate, entire to deeply toothed blades. Flowers are borne in conspicuously showy heads with radiating, bright yellow, three-lobed ray florets and brownish red disk florets subtended by setae. The phyllaries subtending the florets are reflexed. When mature, the turbinate achenes bear five to ten awned scales. Typically encountered as solitary plants or small populations in sandy soils of prairies and open woods, this species occurs in the south-central part of the Great Plains and tallgrass prairie from south-central Kansas south to the Texas Gulf Coast. Use by Native Americans has not been reported, nor has use by wildlife. The binomials *G. lanceolata* and *G. fastigiata* appear in the older literature.

Head of *Gaillardia aestivalis*.

## *Gaillardia pulchella* Foug.

INDIAN BLANKET, FIREWHEEL, ROSE-RING GAILLARDIA

**black, green, purple, yellow**

A member of the Asteraceae (sunflower family), *Gaillardia pulchella* is an annual or rarely biennial herb 10–60 cm tall that arises from a taproot. Branched above, its erect stems bear alternate, simple, sessile or short petioled leaves. The oblanceolate blades are entire, toothed, or rarely pinnatifid. Flowers are borne in conspicuously showy heads with radiating, bright purplish red and yellow, three-lobed ray florets and brownish red disk florets subtended by setae. The phyllaries subtending the florets are reflexed. When mature, the turbinate achenes bear five to ten awned scales. Often forming large populations, this species occupies a variety of soil types and is found in pastures, fields, and roadside rights-of-way. It increases in abundance with heavy grazing and is a dominant of the early to mid stages of plant succession. It occurs throughout the central United States from Colorado, Nebraska, and Missouri south to New Mexico, northern Mexico, and Texas. Populations also extend east along the coast to Florida and Virginia. Flowering is from late May to August. *Gaillardia pulchella* is Oklahoma's state wildflower. Kiowas collected it and took it into their houses as a "good-luck plant." Use of the achenes or foliage by wildlife has not been reported. Taxonomists now consider the species to include taxa formally recognized as *G. drummondii*, *G. villosa*, *G. picta*, and *G. neomexicana*.

Head of *Gaillardia pulchella*.
Photo by Paul Buck.

## ***Glandularia canadensis*** (L.) Nutt.

ROSE VERVAIN

**brown, green, yellow**

A member of the Verbenaceae (vervain family), *Glandularia canadensis* is a perennial herb 30–50 cm tall, with multiple decumbent to ascending stems. Its opposite, simple leaves are incised-pinnatifid or three-cleft. The stems, leaves, bracts, and calyces are hirsute, with glandular and nonglandular hairs intermixed. Borne in dense peduncled spikes 1–1.5 cm in diameter, the flowers have slightly bilaterally symmetrical corollas with pink, rose, blue, lavender, or purple fused petals. The fruits are a quartet of gray-black to black nutlets.

Flowers of *Glandularia canadensis*. Photo by Ronald E. Masters.

Typically encountered in periodically disturbed sites such as roadsides, rights-of-way, and waste places or on rocky prairie hillsides and in open woodlands, this species occurs primarily in the southeastern corner of the United States from North Carolina and Florida west to the eastern third of Kansas, the eastern half of Oklahoma, and the eastern third of Texas. Flowering is primarily from March to May. This species has long been classified as *Verbena canadensis* but was reclassified when *Verbena* was divided into two genera. Although its specific use by Native Americans has not been reported, the leaves of other members of the genus have been used to treat snakebites and sore throats. Several bird species eat this plant's nutlets, but its value for wildlife is minimal.

## ***Gleditsia triacanthos*** L.

HONEY LOCUST

**yellow**

A member of the Caesalpiniaceae (caesalpinia family), *Gleditsia triacanthos* is a polygamous medium tree that readily sprouts from stumps or roots to form tall shrubs and thickets. Its gray-brown bark is fissured into long scaly plates, and its branches have short spurs on the older wood and typically bear stout, shiny, dark brown, branched thorns. The leaves are both one- and two-pinnately compound, with the former fascicled on the spurs and the latter borne singly on the current season's wood. There are nine to four-

Trunk and thorns of *Gleditsia triacanthos*. Photo by Bruce W. Hoagland.

teen pairs of elliptic or oblong-lanceolate leaflets per leaf or pinna. Appearing in clusters of racemes borne on the spurs, its numerous flowers are small, perfect or imperfect, and have radially symmetrical corollas with three to five free, yellowish green petals. At maturity the legumes are oblong, flat, typically twisted, dark gray-brown, stipitate, indehiscent, and 7–45 cm long. Brilliant yellow autumn foliage is also diagnostic. Plants are typically encountered as solitary trees or occasionally in thickets in the moist soils of bottomlands but are also found in drier upland sites. This species is characteristic of the early and mid stages of plant succession but with some individuals persisting in later seres. Flowering is from April to June. The legumes mature in the fall and typically drop en masse. *Gleditsia triacanthos* was originally distributed in the Midwest from the Appalachian Mountains to the Great Plains. It has been planted elsewhere, escaped, and naturalized. Widely planted for shade and in shelterbelts, it is also aggressive in establishment and persistent, and can thus be a pest in introduced pastures and prairies. Native Americans and settlers used the thorns as awls. The sweetish pulp surrounding the seeds in the legumes is edible and eaten by cattle and many species of wildlife, including white-tailed deer and squirrels. Horticulturalists have developed several cultivars. Forma *inermis* lacks thorns. In older references the genus is positioned in the subfamily Caesalpinioideae of the broadly circumscribed family Fabaceae (Leguminosae), the pea family.

### ***Gymnocladus dioicus*** (L.) K. Koch

KENTUCKY COFFEE-TREE

**brown, green, yellow**

A member of the Caesalpiniaceae (caesalpinia family), *Gymnocladus dioicus* is a medium to large polygamodioecious tree up to 30 m tall, with a dark gray, deeply fissured bark. Its stout branchlets have salmon-colored pith and green sapwood. They bear large, alternate, two-pinnately compound leaves. Three to seven pairs of pinnae are present, each bearing four to seven pairs of ovate to elliptic, entire leaflets. The inconspicuous greenish white flowers are produced in terminal panicles and give rise to massive, thick-walled, brown legumes that persist through winter. The large, brown, hard seeds are embedded in a glutinous pulp and must be scarified before they will germinate. This species occurs as solitary trees or small stands in wooded bottomlands, moist ravines, and on lower slopes. Its distribution is primarily the Ohio and Mississippi river drainages as far west as southeastern South Dakota and central Oklahoma. Native Americans used the roots and bark medicinally to treat a variety of problems, particularly to treat constipation and induce sneezing. They and early settlers also ate the roasted seeds and made them into a coffee, hence the common name. Both the foliage and seeds are toxic, producing gastrointestinal irritation and narcotic effects. The toxicant, however, is heat-labile, thus the lack of toxicity after roasting. The wood of this species was once used for cabinets, novelties, fence posts, furniture, and railroad ties. Today it is occasionally planted as an ornamental. Taxonomists differ in their spelling of the specific epithet—the feminine *dioica* versus the masculine *dioicus*. In older references the genus is positioned in the subfamily Caesalpinioideae of the broadly circumscribed family Fabaceae (Leguminosae), the pea family.

Staminate flowers of *Gymnocladus dioicus*. Photo by Paul Buck.

### ***Hedyotis nigricans*** (Lam.) Fosberg

PRAIRIE BLUETS, NARROW-LEAF BLUETS, STAR-VIOLET, FINE-LEAF BLUETS

**brown, orange, yellow**

A member of the Rubiaceae (madder family), *Hedyotis nigricans* is a multistemmed, perennial herb 10–40 cm tall that arises from a branching woody rootstock. Its numerous opposite, simple leaves are sessile, linear, one-nerved, and 20–30 mm long. Borne in crowded, terminal, hemispherical to flat-topped panicles, the small flowers have radially symmetrical corollas of four white to lavender-pink, fused petals. Most conspicuous is the floral dimorphism. The

Flowering plant of *Hedyotis nigricans*. Photo by Dan Tenaglia.

flowers of some plants have four exserted stamens with blue anthers and a style hidden in the corolla throat, whereas flowers of other plants have the four stamens hidden in the corolla throat and the style exserted beyond the corolla. The fruits are small capsules. Upon drying, this plant generally turns black. It typically forms small populations in dry, rocky, shallow soils in prairies and is distributed primarily in the Midwest from southern Michigan, Ohio, and Indiana south to Florida and northern Mexico. Use by Native Americans is not reported, nor is use by wildlife. Taxonomists differ in their interpretation of the boundaries of the genus *Hedyotis*, and the binomial *Houstonia nigricans* is used for this species in some publications. Monographic work conducted by Terrell (1996), however, indicates that prairie bluets should be classified in the genus *Hedyotis*.

### ***Helenium amarum*** (Raf.) H. Rock

BITTER SNEEZEWEED, NARROW-LEAVED SNEEZEWEED, BITTERWEED, YELLOWDICKS

**brown, green, yellow**

A member of the Asteraceae (sunflower family), *Helenium amarum* is an annual herb 10–50 cm tall from a taproot. Its herbage is light green, resinous-punctate, and spicy-smelling when crushed. Its erect stems are profusely branched above, giving the plant a bushy appearance. Alternate or appearing fascicled, the simple leaves are sessile with linear or linear-filiform blades. Flowers are borne in numerous heads at the ends of elongate peduncles. Conspicuously globose, the heads have yellow ray and disk florets with sixteen phyllaries that are reflexed in fruit. The eight to ten ray florets are three-lobed, reflexed, and subtend the numerous disk florets. When mature, the obovoid, four- or five-angled achenes have a pappus of five to eight hyaline scales with apical awns. Growing in a variety of soil types, *H. amarum* is a dominant of the early stages of plant succession and a classic indicator of disturbed soil and overgrazing. Often forming large populations, it is found in waste areas, roadside rights-of-way, and abused pastures. It is distributed in the eastern half of the United States, with greatest abundance in the southeastern

Flowering plant of *Helenium amarum*.

quarter, extending west to eastern Kansas, Oklahoma, and Texas. Flowering is from July to October. Plants of this species are normally unpalatable, but livestock may eat them in the absence of preferred forage and develop a taste for them. The milk of dairy cows that eat this plant will be tainted, generally in the fall. Producing sesquiterpene lactones, *H. amarum* is toxic via inactivation of essential metabolic enzymes. Native Americans and early settlers sniffed crushed leaves to clear the nasal passages and to aid in expulsion of the placenta in childbirth. This plant is considered a good pollen source for bees, although the honey is reputedly bitter. This species may provide screening cover for bobwhite quail, other ground-foraging birds, and small mammals. Consumption of its foliage and achenes by wildlife has not been reported; however, low to moderate use of other species of the genus by white-tailed deer has been documented.

## *Helianthus annuus* L.

ANNUAL SUNFLOWER

**black, brown, green, yellow**

A member of the Asteraceae (sunflower family), *Helianthus annuus* is a tall, scabrous-hispid, annual herb that grows 60–300 cm tall from large taproots. Its leaves are alternate (lowermost opposite), simple, and long-petioled, with ovate to cordate blades. The flowers are borne in large showy heads at the ends of elongate peduncles. The numerous ray florets are yellow, lanceolate, and neutral, whereas the numerous disk florets are brownish purple, perfect, and subtended by receptacular bracts. The obovoid, flattened achenes have a pappus of two lanceolate awns and sometimes one or two scales. Growing in a variety of soil types, this species is a dominant of the early stages of plant succession and a classic indicator of disturbed soil. Allelopathic and often forming large populations, it is found in waste areas, rights-of-way, and at the edges of plowed fields. It occurs across North America with greatest abundance in the Midwest. Flowering is from July to late September. Native Americans used *H. annuus* for food, medicine, fiber, dyeing, and cere-

Heads of *Helianthus annuus*.
Photo by George M. Diggs Jr., courtesy of the Botanical Research Institute of Texas.

monial items. At one time the roasted fruit shells were used to make coffee. Now this plant is commercially cultivated for its oil and achenes or seeds. It is also grown for the cut-flower market and is a popular garden ornamental—the cultivated giant-headed form is a genetic double-recessive trait derived through artificial selection. Although palatable and readily sought out by livestock, this species has low nutritional value. Plants are eaten, to a limited extent, by pronghorn antelope, mule deer, and white-tailed deer. The achenes are an important energy source for many birds and small mammals.

### *Helianthus maximiliani* Schrad.

MAXIMILIAN'S SUNFLOWER

**black, brown, green, yellow**

A member of the Asteraceae (sunflower family), *Helianthus maximiliani* is a robust, perennial herb that arises from thickened rootstocks. Its erect or ascending, solitary or clustered stems are 50–250 cm tall and bear mostly alternate, simple, sessile or short-petioled leaves. The blades are lanceolate, typically folded along the midribs, and have scabrous-hispidulous surfaces. Flowers are borne in large showy heads on long or short peduncles arising in the axils of the uppermost leaves. The ten to twenty-five ray florets are yellow, elliptic to lanceolate, and neutral. The numerous disk florets are likewise yellow and are subtended by receptacular bracts that fold around them. A pappus of two lanceolate awns and sometimes one or two scales crowns the obovoid, flattened achene. Growing in a variety of soil types, this species is commonly encountered in moist sites in prairies and natural drainage ways. Large populations are frequently established in borrow ditches of highway rights-of-way. *Helianthus maximiliani* is characteristic of the late stages of plant succession in prairies and occurs primarily in the center of the continent from southern Manitoba and Saskatchewan to Texas. Populations pre-

Inflorescence of *Helianthus maximiliani*.

sumed to be introductions are found in the Atlantic Coast states. Flowering is from late August to October. Native Americans used both the achenes and rootstocks for food. A prolific achene producer, *H. maximiliani* is a valuable food source for a variety of birds and small mammals. It is browsed by cattle and somewhat by white-tailed deer. Because of its robust growth and profusion of large colorful heads, it is becoming an increasingly popular garden ornamental. Taxonomists differ in their spelling of its specific epithet, some using *maximilianii*.

### ***Helianthus mollis*** Lam.

ASHY SUNFLOWER, HAIRY SUNFLOWER, DOWNY SUNFLOWER

**green, yellow**

A member of the Asteraceae (sunflower family), *Helianthus mollis* is a rhizomatous, perennial herb 50–120 cm tall that often forms small clones. Its villous to hirsute herbage is whitish green. The opposite, simple, sessile leaves have ovate to lanceolate blades with cordate or clasping bases. Flowers are borne in large heads on long or short peduncles arising in axils of uppermost leaves. The five to thirty-five ray florets are yellow, elliptic to lanceolate, and neutral. Subtended by receptacular bracts, the disk florets are numerous and yellow. The obovoid, flattened achenes have a pappus of two lanceolate awns and are villous at the apex. Commonly encountered in sandy or sandy-loam soils, this species occurs in prairies and openings in woods on dry sites. It is characteristic of the late stages of plant succession. It occurs primarily in the Midwest from Ohio and Wisconsin west to eastern Nebraska, Kansas, Oklahoma, and Texas. Populations presumed to be adventive are found east of here as well. Flowering is from July to September. Specific use by Native Americans has not been reported. Like those of other members of the genus, the achenes of *H. mollis* are readily eaten by bobwhite quail, wild turkeys, and other songbirds. The clones also provide cover and are browsed by cattle. Unlike other species of the genus, however, this plant does not appear to be eaten by white-tailed deer.

### ***Helianthus tuberosus*** L.

JERUSALEM ARTICHOKE

**black, brown, green, orange, yellow**

A member of the Asteraceae (sunflower family), *Helianthus tuberosus* is a rhizomatous, perennial herb 1–3 m tall. The rhizomes are tuberous. Opposite below and alternate above, the simple leaves are broadly lanceolate to ovate, usually serrate, distinctly scabrous above, and have winged petioles. Flowers are borne in several to numerous heads on branched peduncles arising

in the axils of the uppermost leaves. The ten to twenty ray florets are yellow, elliptic to lanceolate, and neutral. Subtended by receptacular bracts, the disk florets are numerous and likewise yellow. The obovoid, flattened achenes have a pappus of two lanceolate awns and are glabrous. Typically encountered in both open and shaded, moist to dry sites, this species is distributed throughout the eastern half of the continent and reaches the Great Plains in the west. Native Americans and early settlers ate the tubers in a variety of ways, both raw and cooked, and they remain a popular food item. Like those of other sunflowers, the achenes of *H. tuberosus* are consumed by a variety of wildlife species.

Heads of *Helianthus tuberosus*.
Photo by David G. Smith.

## *Hibiscus laevis* All.

SCARLET ROSE MALLOW, HALBERD-LEAF ROSE MALLOW
**brown, yellow**

A member of the Malvaceae (mallow family), *Hibiscus laevis* is a perennial herb 1–2.5 m tall. Its glabrous stems are typically magenta-tinged and bear alternate, simple, long-petioled leaves with large hastate blades. Borne on long petioles arising in the upper leaf axils, the large showy flowers are subtended by linear-setaceous bracts and have obovate, pink or white petals with dark purple to magenta bases. As is characteristic of the family, the stamen filaments are fused into a column around the

Flower of *Hibiscus laevis*. Photo by Robert J. O'Kennon, courtesy of the Botanical Research Institute of Texas.

style, which ends in five capitate stigmas. Typically encountered in marshes, at pond margins, and in low areas with periodically standing water, this species is distributed throughout the eastern half of the continent, with greatest abundance in the south. Its use by Native Americans and early settlers is not reported. Other members of the genus are popular ornamentals. The binomial *H. militaris* appears in older publications.

### *Hydrangea arborescens* L.

WILD HYDRANGEA

**black, green, yellow**

A member of the Hydrangeaceae (hydrangea family), *Hydrangea arborescens* is a straggly, deciduous shrub 1–3 m tall. Its pale brown bark exfoliates on older branches, and its opposite, simple leaves have ovate to suborbicular, serrate blades with acuminate apices and cordate bases. Borne in a flat-topped or broad hemispheric inflorescence, the flowers are of two types: small, inconspicuous, perfect, fertile flowers are clustered in the inflorescence center, whereas the inflorescence margins are large, showy, white, sterile flowers consisting only of a calyx of three or four fused sepals. The inferior ovaries produce cup-shaped capsules. Characteristically encountered in rocky soils on wooded slopes or along streams in wooded bottoms, this species is distributed from New York and Massachusetts west to eastern Kansas and Oklahoma and south to Florida and Louisiana. Several Native

Flowering plant of *Hydrangea arborescens*. Photo by Paul Buck.

American tribes used its bark medicinally to treat a variety of ailments and in times of scarce food ate its peeled branches and twigs. At one time its root bark was sold as "gravel root," which reflects its use in the treatment of kidney stones (Foster and Duke 2000).

### *Ipomoea leptophylla* Torr.

BUSH MORNING-GLORY

**black, brown, green, yellow**

A member of the Convolvulaceae (morning-glory family), *Ipomoea leptophylla* is a perennial herb 30–120 cm tall. Its multiple, decumbent to erect, branched stems arise from a large rootstock and bear alternate, simple leaves with linear to lanceolate, entire blades and acute apices. Borne at the ends of short axillary peduncles, the radially symmetrical five-merous flowers are solitary or in small clusters. Their funnelform corollas are 5–9 cm long, purple-red to lavender-pink, and dark-throated. In fruit, ovoid capsules contain one to four seeds. Characteristically encountered in sandy prairies or clayey soils of disturbed sites, this species is distributed in the Great Plains from Montana and the Dakotas south to Texas and New Mexico. Plains tribes used its roots as an analgesic, sedative, and stimulant. In times of scarce food, the roots were roasted and eaten. They were also eaten raw to cure stomach troubles and fed to horses to promote fertility and growth of colts.

Flowers of *Ipomoea leptophylla*.
Photo by Bruce W. Hoagland.

### *Ipomopsis rubra* (L.) Wherry

STANDING CYPRESS, TEXAS PLUME, RED GILIA, INDIAN PLUME

**brown, green, orange, red, yellow**

A member of the Polemoniaceae (phlox family), *Ipomopsis rubra* is a biennial herb 1–2 m tall. Its erect, unbranched stems arise from a prominent rosette of leaves and a large taproot. The cauline leaves are alternate and pinnately dissected into ten to fifteen linear to filiform lobes with firm, somewhat sharp apices. Five-merous, radially symmetrical flowers are bright red and salverform with a tube 2–2.5 cm long and are borne singly or in small clusters in the axils of bracts in terminal inflorescences. In fruit, the large oblong capsules contain twenty to twenty-four seeds. Growing in full sun or partial shade and in dry rocky or sandy soils, this species is distributed primarily from Texas and southern Oklahoma east to the Carolinas, Georgia, and Florida. With its showy red flowers that are visited by hummingbirds, this plant is prized as an ornamental; however, it is difficult to maintain a population from year to year. Its specific use by Native Americans has not been reported, but other species of the genus were used by numerous western tribes for medicinal treatments.

Flowering plant of *Ipomopsis rubra*.
Photo by George M. Diggs Jr., courtesy of the Botanical Research Institute of Texas.

## *Juglans nigra* L.

BLACK WALNUT

**brown, orange**

A member of the Juglandaceae (walnut family), *Juglans nigra* is a large monoecious tree up to 40 m tall. Its dark gray to brown bark is deeply furrowed into narrow ridges, and the stout branchlets bear alternate one-pinnately compound leaves with fifteen to nineteen lanceolate yellow-green leaflets. The terminal leaflet is absent. Its heart-shaped or three-lobed leaf scars are diagnostic, as are its chambered pith and its spherical nut enclosed in a large seamless involucral husk. Characteristic of bottomlands throughout the eastern half of the United States, this species is of considerable economic importance. Its wood is used to make furniture, pan-

Leaves and fruits of *Juglans nigra*.

eling, gunstocks, and bowls, and its nuts are prized as food and flavoring. Native Americans used decoctions of the bark as emetics, infusions of the leaves as washes for sores, and applications of the sap and husk juice as treatment for inflammation and ringworm. Sensitive individuals may develop contact dermatitis after handling the leaves or fruit husks. Laminitis in horses is a problem if the animals are bedded on fresh shavings of the wood.

## *Juniperus virginiana* L.

EASTERN RED CEDAR

**black, brown, green, orange, red, yellow**

A member of the Cupressaceae (cypress family), *Juniperus virginiana* is a strongly aromatic small to medium tree up to 30 m tall. It is readily recognized by its pyramidal evergreen growth form, its thin reddish brown bark that shreds into long narrow strips, its four-ranked scale-like leaves, and its glaucous blue seed cones or "berries." This relatively slow-growing species occupies a variety of soils and habitats, including rocky dry uplands and swampy bottomlands, throughout the eastern half of the United States west to the Dakotas, Nebraska, Kansas, Oklahoma, and Texas. In prairie regions it is believed to have been confined originally in distribution to shallow soils, rocky ridges, and bluffs where intense fires did not occur. It does not sprout after top-kill. Suppression of fire in prairies, introduced pastures, old fields, riparian areas, forests, shrublands, and other plant communities has allowed this species to spread. Pollination occurs in April and May. Native Americans made an array of medicines from all parts of *J. virginiana* to cure mouth sores, head colds, coughs, kidney problems, nervous

Foliage and seed cones of *Juniperus virginiana*. Photo by George M. Diggs Jr.

problems, and other ailments. It was important in purification rituals as well. Settlers used the species extensively to make fence posts, roof rafters, and windowsills. The aromatic wood repels insects, especially moths, and is used to line chests and closets. Trees are frequently planted in shelterbelts. The major drawback to its use is its ability to invade other communities. Because of its evergreen habit, *J. virginiana* provides cover, especially in winter, for birds and small mammals. Its seed cones are rich in carbohydrates and are eaten by numerous wildlife species.

## ***Lesquerella gracilis*** (Hook.) S. Watson

SPREADING BLADDERPOD

**black, yellow**

A member of the Brassicaceae (mustard family), *Lesquerella gracilis* is an annual or biennial herb with both decumbent and erect stems that are 10–30 cm tall. Its young stems and leaves are gray-green with stellate hairs, whereas the older parts are sparsely pubescent and green. The alternate, simple, sessile or subsessile, dentate leaves are oblanceolate to obovate. Borne in terminal racemes, the four-merous, radially symmetrical flowers have bright yellow to orange, broadly obovate petals and six stamens (four long and two short). In fruit, the globose to ellipsoidal silicles split apart, leaving a central partition. Growing in sandy, clayey, and heavy black soils of prairies, open fields, and disturbed roadsides, this species is distributed primarily in the lower Midwest from Illinois, Missouri, and Nebraska south to Alabama and Texas. Although its specific use by Native Americans has not been reported, the roots and leaves of other species of the genus were used medicinally to treat diarrhea, swellings, heartburn, sores, toothaches, and sore eyes. Taxonomists recognize two varieties on the basis of the appearance of the silicle and the toothing of the leaves.

Flower of *Lesquerella gracilis*.
Photo by George M. Diggs Jr.

## ***Liatris punctata*** Hook.

DOTTED GAYFEATHER, DOTTED BLAZING-STAR

**black, brown, green, yellow**

A member of the Asteraceae (sunflower family), *Liatris punctata* is a perennial herb with erect or ascending, generally unbranched, solitary or clustered stems that are 15–50 cm tall. It is readily recognized by its massive, elongate, fibrous rootstock; its linear, punctate, sessile leaves; and its dense terminal spikes of elongate heads bearing only disk florets. There are four to eight disk florets per head, and their corollas are purple-lavender. A pappus of plumose capillary bristles is at the top of a ten-ribbed achene. Occupying a variety of clayey, loamy, and sandy soil types, this species occurs in the mid and late stages of plant succession in the mixed and tallgrass prairies, from Alberta, Saskatchewan, and Manitoba south to Texas and northeastern Mexico. It increases with heavy grazing and may form large populations. Plants also occur on sites where the soil has been disturbed. Flowering is from August until October. Native Americans and early settlers used various parts of the plants to treat a variety of ailments, including stomachaches, swellings, urinary problems, and gastrointestinal complaints. Pronghorn antelope graze its herbage, but its small achenes are not consumed. This plant is quite showy, with its striking inflorescences of purple-lavender heads, and is frequently cultivated by individuals creating native plant gardens.

Heads of *Liatris punctata*.

### ***Liatris squarrosa*** (L.) Michx.

SCALY GAYFEATHER, SCALY BLAZING-STAR

**green, yellow**

A member of the Asteraceae (sunflower family), *Liatris squarrosa* is a perennial herb with erect, generally unbranched stems 30–60 cm tall that arise from a prominent rootstock. The flowers are borne in showy campanulate heads containing only disk florets. There are twenty-five to forty florets per head, and their corollas are purple-lavender. The mature achene bears a pappus of plumose capillary bristles. This species typically grows in sandy or calcareous soils

Heads of *Liatris squarrosa*. Photo by George M. Diggs Jr., courtesy of the Botanical Research Institute of Texas.

of dry upland prairies and open woods and is distributed from Maryland to Florida and west to South Dakota and Texas. Flowering is from July to September. Although its specific use by Native Americans has not been reported, the roots of other species of the genus were used medicinally to treat a variety of ailments, including stomachaches, swellings, urinary problems, and gastrointestinal complaints.

## *Linum sulcatum* Riddell

GROOVED FLAX, GROOVED YELLOW FLAX

**brown, green, yellow**

A member of the Linaceae (flax family), *Linum sulcatum* is an erect, glabrous, annual herb 20–80 cm tall. Arising from a taproot, its erect stems are typically branched above and bear simple leaves that are alternate above and opposite below. Their entire blades are linear to narrowly lanceolate, with sharp-pointed apices. Borne in open inflorescences, the five-merous, radially symmetrical flowers have free, pale yellow petals that are pubescent at their bases. At maturity the capsule dehisces into ten one-seeded, sharp-pointed segments. Typically growing in sandy soils of prairies and open woodlands, this species is distributed throughout the eastern half of the continent but is more common in the most western portion of its range. Flowering occurs from May to September. This plant's specific use by Native Americans has not been reported, but other species of the genus were used medicinally to make infusions, decoctions, and poultices. Other species were also used for food and as a source of fiber.

Flower of *Linum sulcatum*.

## *Liquidambar styraciflua* L.

SWEET GUM, RED GUM

**black, brown, green, orange, red, yellow**

A member of the Hamamelidaceae (witch hazel family), *Liquidambar styraciflua* is a large monoecious tree up to 40 m tall. Its bark is gray-brown and deeply furrowed, and its multiwinged branchlets bear alternate, simple, palmately five-lobed leaves that are fragrant when crushed. Its imperfect flowers lack sepals and petals. The staminate flowers are borne in dense green racemes, whereas the pistillate flowers are borne in globose heads at the ends

Leaves of *Liquidambar styraciflua*. Photo by Steven K. Goldsmith.

of long pendulous peduncles. In fruit, the syncarps of spiny capsules persist on the tree. Characteristically encountered in floodplains, swamps, moist slopes, and low-lying fields, this species is distributed primarily in the southeastern quarter of the United States. It is somewhat weedy and may form dense populations in highway borrow ditches, moist disturbed areas, and abandoned fields. It is pathogen-resistant and susceptible to fire. Flowering is in April and May. As the common name "sweet gum" implies, *L. styraciflua* is aromatic: its leaves are quite fragrant when crushed, as is the wood when cut. Gum from the bark has been used in skin care products, adhesives, pipe tobacco, and perfume. Native Americans used the gum, bark, and roots to treat a variety of ailments, including diarrhea, wounds, sores, and nervousness. This plant has brilliant fall color and is a popular ornamental, with a number of cultivars.

### *Maclura pomifera* (Raf.) C.K. Schneid.

OSAGE ORANGE, HORSEAPPLE, HEDGEAPPLE, BOW-WOOD, BODARK, BOIS D'ARC
**black, brown, green, yellow**

A member of the Moraceae (mulberry family), *Maclura pomifera* is a small to medium tree, often with several trunks, that is readily recognized by its orange, papery inner bark; its milky sap; its stout thorns; its glossy, dark green leaves; and its large, spherical, yellow-green fruits. Its original distribution is reputed to have been the Red River drainage in Oklahoma and Texas. Because of its many uses, it has been planted across the continent, and in many areas it has escaped and naturalized. This fast-growing species thrives in a variety of soils and habitats, including prairies, open woody uplands, and bottomlands. It is characteristic of the mid stages of plant succession. Aggressive in establishment and persistent, it is capable of invading native prairies, introduced pastures, and old fields. It is a vigorous root sprouter, often forming dense hedgerows. Flowering is in May. In the fall, the large fruits litter the ground beneath the pistillate trees. *Maclura pomifera* has long been used for windbreaks, hedgerows, woodlots, and shelterbelts. Its extremely hard wood is prized for fence posts and was even used for road paving in Sherman, Texas, in the 1800s. It is among the best fireplace woods, too: a cord will produce 28.3 million Btus. The aromatic fruits repel insects in closets and cupboards. The milky sap may cause contact dermatitis in some individuals. Native Americans used the wood for bows, hence common names such as "bow-wood." It was also used to make war clubs and was used for dyeing and tanning. Cattle and white-tailed deer browse the foliage and twigs to a limited extent. Small mammals tear apart the fruits to eat the achenes, and birds eat the remnants.

Leaves and fruit of *Maclura pomifera*.

## ***Mentzelia nuda*** (Pursh) Torr. & A. Gray

BRACTLESS BLAZING-STAR, SAND LILY, POOR MAN'S PATCHES, STAR-FLOWER

**black, brown, green, yellow**

A member of the Loasaceae (stick-leaf family), *Mentzelia nuda* is a biennial or perennial herb that arises from a stout taproot. Its erect, branched stems are up to 100 cm tall and bear alternate or subopposite, simple leaves with sessile, linear-lanceolate, bluntly dentate blades and scabrous surfaces. Borne near the branch tips and typically open from mid afternoon until sunset, the showy, radially symmetrical flowers have ten to twelve white to cream petals that arise from an inferior ovary. In fruit, the cylindrical capsules are topped by five persistent, lanceolate sepals. Typically encountered in sandy or gravelly soils in sandy prairies or on eroded banks as scattered plants or localized populations, this species is distributed in the Great Plains and Rocky Mountains from Canada south to Texas, New Mexico, and Arizona. Flowering occurs from July through September, and the appearance of the large showy flowers in the late afternoon and evening is quite striking. Native Americans boiled and strained its sap to make a skin wash to treat fevers. The binomial *M. stricta* is encountered in older publications. Today many taxonomists recognize the varieties *nuda* and *stricta*. The morphologically similar *M. decapetala* differs by having larger flowers and overlapping petals.

Flowers of *Mentzelia nuda*.
Photo by Bruce W. Hoagland.

## ***Monarda fistulosa*** L.

WILD BERGAMOT, LONG-FLOWERED HORSEMINT, PURPLE BEEBALM

**brown, green, yellow**

A member of the Lamiaceae (mint family), *Monarda fistulosa* is an aromatic, perennial herb 70–120 cm tall that often forms dense clumps or small colonies from slender rhizomes. Its erect stems are four-sided and bear opposite, simple leaves with ovate to lanceolate, serrate to entire blades and glandular-punctate surfaces. The showy inflorescence is a large, solitary verticil of flowers at the stem apex. As is characteristic of the

Inflorescence of *Monarda fistulosa*.
Photo by Wayne J. Elisens.

mints, the bilaterally symmetrical flowers are showy and perfect, with pale to dark lavender, two-lipped corollas. In fruit, four brown or black nutlets are present. Typically encountered in loamy soils at forest edges, on prairie hillsides, and in wet meadows, marshes, and ditches, this species occurs across the continent except in Florida, Nevada, and California. Flowering is from May to September. Native American tribes used parts of *M. fistulosa* in various ways. It was used as a medicine to treat colds, fevers, flu, coughs, wounds, sores, headaches, bronchitis, and gastrointestinal problems; as a food, seasoning, and food preservative; as an ingredient in perfumes; and as a fragrance for rooms and bedding. It was also used to make cooking tools.

## *Morus rubra* L.

RED MULBERRY

**brown, green, yellow**

A member of the Moraceae (mulberry family), *Morus rubra* is a small to large, typically dioecious tree up to 20 m tall. Its dark gray-brown bark is scaly or shallowly furrowed. When cut, its red-brown or green-brown branchlets exude a milky sap. The alternate, simple leaves have thin, ovate, serrate blades with acuminate apices. Typically the leaves have a mitten-like appearance due to the presence of one large lobe. Borne in short cylindrical catkins, the staminate and pistillate flowers are inconspicuous. In fruit, however, the multiple syncarps of fleshy calyces partially or entirely enclosing the achenes are dark purple. This species is an understory tree and is generally encountered in shaded bottomlands or moist slopes. It is distributed throughout the eastern half of the United States from New York and Massachusetts west to eastern Nebraska and south to Florida and central Texas. Flowering is in April and May, with the fruits persisting until July. Mulberries are relished both by birds and humans and are readily eaten. They are also used in pies, jellies, and jams. Native Americans used the species medicinally. It is considered an excellent species for wildlife, and its wood is used for fence posts, furniture framing, and interior trimwood. It is a popular ornamental, though not as widely grown as the introduced *M. alba* (white mulberry).

Leaves of *Morus rubra*.

### ***Neptunia lutea*** (Leavenw.) Benth.

YELLOW NEPTUNE, YELLOW PUFF

**black, brown, yellow**

A member of the Mimosaceae (mimosa family), *Neptunia lutea* is a perennial herb that arises from a woody taproot. It is readily recognized by its two-pinnately compound leaves bearing numerous small leaflets that fold when touched, and by its globose spikes of yellow, radially symmetrical flowers with their conspicuous exserted stamens. It is also characterized by its prostrate habit and prickle-free stems. Characteristic of mid to late stages of plant succession, this species typically grows in dry rocky sites of prairies, open woodlands, and disturbed soils of roadsides. It is generally not abundant, but it does form small scattered populations. It occurs in the south-central portion of the United States from Alabama and Mississippi west to Texas, Oklahoma, and southern Kansas. Flowering occurs from April to October. Use by Native Americans is not reported. The herbage of *N. lutea* is occasionally eaten by white-tailed deer and cattle, and the seeds are consumed to a limited extent by bobwhite quail, wild turkeys, and other wildlife species.

Inflorescences and leaves of *Neptunia lutea*. Photo by Robert J. O'Kennon.

### ***Nuttallanthus canadensis*** (L.) D.A. Sutton

TOAD-FLAX, OLDFIELD TOAD-FLAX

**brown, yellow**

A member of the Scrophulariaceae (figwort family), *Nuttallanthus candensis* is a slender, annual or winter-annual herb 10–50 cm tall from a short taproot. Its slender, erect or ascending stems bear al-

Flowers of *Nuttallanthus canadensis*. Photo by Charles S. Lewallen.

ternate, simple leaves that are sometimes opposite at the stem base. Borne in elongate, terminal racemes, the flowers are light blue to lavender with a bilabiate corolla and a slender, curved spur. Its subglobose capsules contain numerous seeds. This species typically occurs as scattered plants or small localized populations in sandy soils of disturbed sites and fallow fields. Its distribution is essentially across the continent with the exception of the Rocky Mountains and Intermountain Region. Flowering is from April to June. Use by Native Americans is not reported. Toad-flax has long been classified as *Linaria canadensis*, but taxonomic work indicates that *Linaria* is heterogeneous with the New World species different from those of the Old World, hence the segregation of the New World taxa into *Nuttallanthus*.

### ***Nyssa sylvatica*** Marshall

BLACK GUM, BLACK TUPELO, SOURGUM

**black, brown, green, orange, yellow**

A member of the Nyssaceae (sourgum family), *Nyssa sylvatica* is a straight, medium to large tree up to 30 m tall. Its dark bark comprises hexagonal plates, and the white pith of its branchlets is partitioned. The alternate, simple leaves have oblong-elliptic, entire blades. As is characteristic of this species, the leaves begin to develop red spots and blotches early in summer and become a brilliant scarlet in autumn. In fruit, clusters of two to four blue-black drupes are borne at the ends of curved peduncles. Typically encountered in bottomlands, freshwater swamps, and on moist lower slopes, this species is distributed in the eastern half of the United States but is most common in the southeastern quarter. It is not a true dominant of the forest canopy. Flowering is in May and June. The wood is used to make furniture, implement handles, poles, railroad ties, and pulp, and in Appalachia is used for beehives. Native Americans made infusions and decoctions of the roots and bark, often mixed with other plants, to treat worms, diar-

Leaves of *Nyssa sylvatica*.

rhea, eye problems, tuberculosis, and problems arising during childbirth. It was also used to induce vomiting. A variety of birds eat its drupes, mammals browse its herbage, and it is considered a good honey tree. Its rapid growth and showy autumn foliage make it a desirable ornamental as well.

### *Oenothera heterophylla* Spach

SAND EVENING-PRIMROSE, VARIABLE EVENING-PRIMROSE

**black, brown, green, orange, red, yellow**

A member of the Onagraceae (evening-primrose family), *Oenothera heterophylla* is a winter-annual or biennial herb that grows up to 100 cm tall from a fleshy taproot. Its alternate, simple leaves have lanceolate, entire or sinuate-dentate blades. The upper leaves are sessile and clasp the stem. The showy four-merous flowers are borne in elongate terminal spikes and open around sunset; they have reflexed sepals and large yellow petals borne at the end of a long tubular hypanthium. The mature capsules dehisce along four sutures. This species typically forms conspicuous colonies in sandy soils at the edges of woods or on roadside banks, occurring in Texas, western Louisiana, southern Oklahoma, and western Arkansas. Flowering is from June to September. This plant's specific use by Native Americans has not been reported, but other species of the genus were used as medicine, food, dye, and in ceremonies.

Flowers and capsules of *Oenothera macrocarpa*.

### *Oenothera macrocarpa* Nutt.

BIGFRUIT EVENING-PRIMROSE, MISSOURI EVENING-PRIMROSE, MISSOURI PRIMROSE

**black, brown, green, yellow**

A member of the Onagraceae (evening-primrose family), *Oenothera macrocarpa* is an acaulescent or caulescent, perennial herb that arises from a woody rootstock. The stems, if present, are ascending or decumbent and bear alternate, simple leaves with elliptic to lanceolate, entire to sparsely serrate blades. The large, showy, four-merous flowers open near sunset and have reflexed sepals and large yellow petals borne at the end of a long tubular hypanthium. Most conspicuous are the large four-winged capsules that turn a brilliant purple-red as they mature. Growing in a variety of soil types on rocky prairie hillsides, limestone escarpments, and eroded roadside banks, this species is distributed from Illinois and Nebraska south to Texas, with populations also occurring in Tennessee. Flowering is from May through July. This plant's use by Native Americans has not been reported, but other species of the genus were used as medicine, food, dye, and in ceremonies. Taxonomists recognize four varieties, based on leaf shape, pubescence, and petal size. The binomial *O. missouriensis* is used in older publications.

## ***Opuntia macrorhiza*** Engelm.

PRICKLY-PEAR, PLAINS PRICKLY-PEAR, BIGROOT PRICKLY-PEAR, TWISTSPINE PRICKLY-PEAR

**yellow**

A member of the Cactaceae (cactus family), *Opuntia macrorhiza* is a succulent, herbaceous or suffrutescent perennial. Its branched and jointed stems are flattened to form pads, and its leaves are modified spines arising from areoles. Large, showy, yellow, solitary flowers also arise from the areoles. In fruit, the obovoid berries are reddish purple and contain numerous seeds. Occupying a variety of soil types, this species is found in dry prairies, where it increases in abundance with heavy grazing and disturbance. If fragmented, the pads produce adventitious roots and establish new plants. *Opuntia macrorhiza* occurs in the Great Plains and tallgrass prairie from South Dakota to southern California, Arizona, New Mexico, Texas, and Louisiana. Populations also reportedly occur as far east as Ohio and Kentucky. Flowering is from May to July. When the areoles are removed, the fleshy fruits are edible and can be eaten raw or added to stews and soups. Native Americans dried them for use in winter. Roasted pads were eaten in times of star-

Flower and pad of *Opuntia macrorhiza*. Photo by Robert J. O'Kennon.

vation by tribes of the Missouri River drainage, and modern-day ranchers sometimes burn the spines off so that cattle can eat the pads when other forage is scarce. Scaled quail, wild turkeys, and small mammals eat the fruits and seeds. Pronghorn antelope and white-tailed deer consume the pads and fruits. When abundant, this plant can be an important component of cover for bobwhite quail.

### ***Packera obovata*** (Muhl. ex Willd.) W.A. Weber & A. Löve

ROUNDLEAF GROUNDSEL, GOLDEN GROUNDSEL, CREEPING GROUNDSEL

**brown, green, yellow**

A member of the Asteraceae (sunflower family), *Packera obovata* is a glabrous, stoloniferous perennial 20–50 cm tall. It is characterized by its basal rosette of long-petiolate, crenate-serrate, obovate or oblanceolate to spathulate leaves. The upper cauline leaves are reduced, sessile, and typically pinnatifid. Borne in terminal corymbose clusters, the cylindrical to campanulate heads have their phyllaries in two distinct series and comprise eight to thirteen yellow ray florets and numerous yellow disk florets. In fruit, the cylindrical five- to ten-ribbed achenes have a pappus of numerous barbellate hairs. This species is typically encountered growing in moist loamy soils on woody hillsides, in intermittent stream beds, and in calcareous rocky outcrops, occurring throughout the eastern part of the continent, and reaching eastern Kansas and Oklahoma, southern Texas, and northern Mexico. Flowering is in April and May. Although use by Native Americans

Flowering plants of *Packera obovata*. Photo by Robert J. O'Kennon.

has not been reported for this species, other members of *Packera* were used medicinally and in ceremonies. Other members of the genus are also known to produce compounds collectively known as pyrrolizidine alkaloids, nitrogen-containing ring compounds that are hepatotoxic and cause problems in livestock; however, *P. obovata* has not been implicated. Roundleaf groundsel has long been known as *Senecio obovatus*. Taxonomic work, however, indicates that *Senecio* is a heterogeneous genus, and so it has been divided into several genera, including *Packera*, with roundleaf groundsel reclassified as *P. obovata*.

## ***Parthenocissus quinquefolia*** (L.) Planch.

VIRGINIA CREEPER, WOODBINE

**black, brown, green, orange, yellow**

A member of the Vitaceae (grape family), *Parthenocissus quinquefolia* is a high-climbing woody vine. It is readily recognized by its branched tendrils with terminal pads; its palmately compound leaves with five oblanceolate to elliptic, serrate leaflets; and its clusters of black or blue-black berries in peduncled cymes. The bright red or scarlet fall foliage is also diagnostic. This species is often confused with the superficially similar *Toxicodendron radicans* (poison ivy), which has pinnately compound leaves, adventitious roots rather than tendrils, and white drupes arising laterally along the stems. Typically found attached to the trunks of trees and forming large masses of foliage in their crowns, *P. quinquefolia* is sometimes herbaceous and spreads

Leaves and fruits of *Parthenocissus quinquefolia*.

by rhizomes across the forest floor, especially following periodic fire. It is characteristic of the mid to late stages of plant succession. Flowering is in May and June. The berries mature in the fall and often persist in early winter if not eaten by wildlife. This species typically occupies forest edges and openings, occurring throughout the eastern half of the United States as far west as southeastern South Dakota and eastern Nebraska, Kansas, Oklahoma, and Texas. Native Americans made infusions, decoctions, and poultices of the leaves, twigs, and bark to treat jaundice, diarrhea, swelling of the joints, and wounds. The fruits were used to make a paint for skin and feathers. Bobwhite quail, many species of songbirds, and small mammals eat the berries. Squirrels eat the bark, foliage, and fruit in winter. White-tailed deer commonly browse the leaves and stems in spring and summer.

## *Pediomelum cuspidatum* (Pursh) Rydb.

TALL-BREAD SCURFPEA, LARGE-BRACT INDIAN-BREAD

**brown, green, yellow**

A member of the Fabaceae (pea family), *Pediomelum cuspidatum* is a perennial herb that arises from a thickened, branched taproot. Its decumbent or ascending stems are 30–60 cm long and bear alternate, palmately compound leaves with five or rarely three elliptic to rhomboidal leaflets. Stout axillary peduncles terminate in elongate racemes. Subtended by ovate to elliptic bracts, the blue or purplish flowers have papilionaceous corollas and typically the pleasant vanilla-like odor of coumarin (a scent similar to that of sweet-clover). The small legumes are one-seeded. This species normally grows in calcareous loamy soils, occurring as scattered plants or small populations in dry prairies, gravelly hilltops, bluffs, and slopes. It is distributed in the Great Plains from Montana and the Dakotas south to Texas. Flowering is from May to July. Native Americans used this species and its relatives medicinally. Its toxicological potential is unclear, but it may contain phototoxic furanocoumarins, as do other members of the genus. The binomial *Psoralea cuspidata* has long been used for this species. Taxo-

Inflorescence of *Pediomelum cuspidatum*.
Photo by Bruce W. Hoagland.

nomic work, however, has revealed that *Psoralea* is strictly an Old World genus characterized by a cupule below each flower. New World species of scurfpea lack this cupule and are therefore classified in the genera *Psoralidium*, *Pediomelum*, *Orbexilum*, and *Hoita*.

## *Phlox pilosa* L.

PRAIRIE PHLOX, DOWNY PHLOX

**black, brown, green, yellow**

A member of the Polemoniaceae (phlox family), *Phlox pilosa* is a perennial herb that grows 30–60 cm tall from a stout rootstock. It is readily recognized by its glandular-pilose herbage; its opposite, sessile, stiff, sharp-pointed leaves; and its dense cymes of flowers with salverform corollas and purple, lavender, pink, or white petals. In fruit, its capsule contains only three winged seeds. Growing primarily in dry sites with thin sandy or rocky-loamy soils, this species is found in both prairies and open woodlands. It is typically encountered as small localized populations and occurs across the eastern half of the continent to Manitoba, the Dakotas, Nebraska, Kansas, Oklahoma, and Texas. Flowering is from April through early June. Native Americans made infusions of the leaves to purify the blood and treat eczema. Cattle may eat the plants in early spring. White-tailed deer readily consume them in spring and moderately so in summer and fall, and prairie chickens and wild turkeys sometimes eat the capsules.

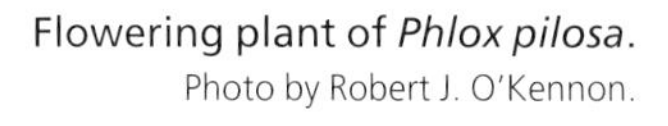

Flowering plant of *Phlox pilosa*.
Photo by Robert J. O'Kennon.

## *Phyla lanceolata* (Michx.) Greene

LANCELEAF FROG-FRUIT, NORTHERN FROG-FRUIT, FOG-FRUIT

**brown, green, yellow**

A member of the Verbenaceae (vervain family), *Phyla lanceolata* is a creeping, perennial herb with prostrate to ascending, somewhat four-sided stems 20–60 cm long that often root at the nodes. Its opposite, simple leaves have

Flowering plant of *Phyla lanceolata*. Photo by Bruce W. Hoagland.

lanceolate to elliptic blades with serrate margins above the middle. Borne in dense, globose to cylindrical spikes on elongate peduncles, the four-merous flowers have a slightly two-lipped, white or pale lavender corolla, often with a yellow center. In fruit, two nutlets remain enclosed in the calyx. Typically encountered as spreading clones in wet or moist soils of floodplains, ditches, muddy flats, pond margins, and prairie swales, this species occurs across the continent, except for the northwest corner. Flowering is from May to September. Mahunas used it to treat rheumatism. Taxonomists differ in their classification of lanceleaf frog-fruit and related species: some place them in *Phyla*, whereas others place them in the segregate genus *Lippia*.

### *Physalis angulata* L.

CUTLEAF GROUND-CHERRY

**yellow**

A member of the Solanaceae (nightshade family), *Physalis angulata* is a glabrous or glabrate, annual herb 20–100 cm tall, with erect or ascending stems that typically branch at the plant base. Its alternate, simple leaves have ovate to ovate-lanceolate, irregularly toothed blades. Solitary five-merous flowers are borne on recurved pedicels in the leaf axils. Their weakly five-lobed corollas are yellowish, and the anthers are bluish or violet. As is characteristic of the genus, the flowers are most conspicuous in fruit, with inflated, ten-angled or ten-ribbed calyces surrounding the berries. Typically encountered

Leaves and flowers of *Physalis angulata*. Photo by Gerald D. Carr.

as scattered plants or small localized populations in a variety of soil types in different habitats, this species occurs throughout the southern half of the continent and extends south to Central America and the West Indies. Flowering is from June to October. This plant's use by Native Americans is not reported, but other species of the genus were used as food and medicine. The berries were eaten raw or cooked and were dried and stored for winter. The berries, similar in appearance to tomatillos, the fruits of another species of *Physalis*, are also relished by game birds and mammals, although they are a minor part of their diets.

### ***Physalis virginiana*** Mill.

VIRGINIA GROUND-CHERRY

**green, orange, yellow**

A member of the Solanaceae (nightshade family), *Physalis virginiana* is a pubescent, rhizomatous, perennial herb 30–60 cm tall, with erect stems that are usually forked with ascending branches. Its alternate, simple leaves have ovate to lanceolate, entire or toothed blades. Solitary five-merous flowers are borne on recurved pedicels in the leaf axils. Their five-lobed corollas are yellowish with dark blotches, and their anthers are yellow. As is characteristic of the genus, the flowers are most conspicuous in fruit, with the inflated, five-angled calyces surrounding the berries. Typically encountered as scattered plants or small localized populations in a variety of soil types in

fields, upland woods, and prairies, this species occurs throughout the eastern two-thirds of the continent, as far west as the flanks of the Rocky Mountains. Flowering is from May to September. Native Americans ate the berries and made infusions of the whole plant to treat dizziness. The berries, similar in appearance to tomatillos, the fruits of another species of *Physalis*, are also relished by game birds and mammals, although they are a minor part of their diets.

Flower of *Physalis virginiana*.
Photo by Charles S. Lewallen.

## *Phytolacca americana* L.

POKEWEED, POKEBERRY, POKE, AMERICAN POKEWEED
**yellow**

A member of the Phytolaccaceae (pokeweed family), *Phytolacca americana* is a glabrous, short-lived perennial herb that grows from fleshy roots. It is readily recognized by its erect 1- to 3-m stems that turn purplish red at maturity; its simple, alternate leaves with ovate to lanceolate, entire to undulate blades; and its long, drooping, axillary racemes of greenish white or cream flowers that produce purplish black berries. Growing in a variety of clayey, loamy, and sandy soils, this species is typically associated with severe disturbance. Often found on logged sites with slash piles or on mounds of soil with uprooted trees, it also occupies roadsides, ditch banks, and waste areas. It is a dominant of the early stages of plant succession and is found in both prairies and open woodlands. Flowering is from June through October. The seeds are dispersed by songbirds and persist in the soil seed bank for forty years or more. It is distributed across the eastern half of the continent to Min-

Fruits of *Phytolacca americana*.
Photo by George M. Diggs Jr.

nesota, Nebraska, Kansas, Oklahoma, and Texas. When properly cooked, the tender shoots and young leaves are edible and rich in vitamin C; otherwise they are poisonous and can cause severe diarrhea. Cans labeled "poke salet" may be found on grocer's shelves, and Euell Gibbons called it a potherb par excellence. The berries are edible when cooked and are often used in pies and jellies. Native Americans used the berries to make teas to treat rheumatism and used the roots to treat skin diseases. In Appalachia the juice of the crushed berries has been drunk to purify the blood and regularly added to the water of chickens to keep them disease-free. Many species of birds and small mammals eat the berries.

## *Platanus occidentalis* L.

SYCAMORE, PLANE-TREE, BUTTONWOOD

**brown, yellow**

A member of the Platanaceae (plane-tree family), *Platanus occidentalis* is a large monoecious tree up to 50 m tall and 4 m in diameter. Its mottled brown-gray bark exfoliates in long plates to reveal smooth greenish white surfaces. The alternate, simple leaves have large three- to five-palmately lobed, coarsely serrate blades. The petioles completely enclose the axillary buds. Its imperfect flowers are borne in globose, axillary heads. In fruit, the pistillate heads comprising many achenes surrounded by stiff hairs are pendulous on long, flexuous peduncles and typically persist until the next growing season. Characteristically encountered in wet soils of stream banks and floodplains, this species is distributed throughout the eastern half of the continent as far west as the Great Plains. Flowering is from March to May. Native Americans made infusions and decoctions of the bark, often combined with other plants, to treat gastrointestinal, urinary, gynecological, skin, and respiratory problems, as well as other ailments. The wood is used to make furniture, millwork, flooring, and butcher blocks. At one time it was used for cheese boxes. Accounts of early explorers in North America describe how they used the hollow trunks of giant old sycamores for shelter during storms. Individuals differ in their ability to smell the sweet resinous odor of the leaves and cut wood.

Leaves of *Platanus occidentalis*.

## *Podophyllum peltatum* L.

MAY-APPLE, AMERICAN MANDRAKE, INDIAN APPLE, UMBRELLA LEAF, PODOPHYLLUM, GROUND LEMON

**black, brown, yellow**

A member of the Berberidaceae (barberry family), *Podophyllum peltatum* is a rhizomatous, perennial herb typically forming extensive vegetative colonies of scattered, peltate, deeply lobed leaves borne on stout, stem-like petioles. When reproducing, stems 30–50 cm tall bearing two petiolate, peltate, lobed leaves appear. In the fork of these leaves is a solitary, large, showy, white or pink flower on a curved pedicel. In fruit, the berries are yellow to orange, red, or maroon, and 4–5 cm in diameter. One of the classic harbingers of spring in eastern deciduous forests, this species is indicative of moist, rich soils of shaded forest floors and is distributed throughout the eastern half of the continent as far west as Minnesota and eastern Nebraska, Kansas, Oklahoma, and Texas. Its medicinal and toxic effects were well known to Native Americans and colonists. It was used as an emetic in the 1700s. The roots have purgative effects and have been officially recognized as a cathartic and cholagogue in the United States Pharmacopeia. In the 1900s, podophyllin, the resin extracted from the roots of this plant, was incorporated into proprietary medicines such as Carter's Little Liver Pills (Graham and Chandler 1990). The berries are edible when ripe but can cause problems if excessive numbers are eaten.

Leaves and flower of *Podophyllum peltatum*. Photo by Steven K. Goldsmith.

## *Polygala alba* Nutt.

WHITE MILKWORT

**brown, yellow**

A member of the Polygalaceae (milkwort family), *Polygala alba* is a perennial herb that arises from a stout, vertical rootstock. Its multiple ascending to erect stems are 20–40 cm tall and bear alternate, simple, linear to oblanceolate leaves. Borne in short, showy spikes at the ends of long peduncles, its

Flowering plant of *Polygala alba*. Photo by Bruce W. Hoagland.

small white or greenish white flowers have five sepals, two of which are petaloid, wing-like, and larger than the three or rarely five petals. The flowers superficially resemble the papilionaceous flowers of the Fabaceae (pea family), but upon close examination it is evident that the structures are not the same. This species grows in a variety of soil types on rocky outcrops and hillsides of prairies and open woods, and in dry washes and cedar thickets, occurring primarily in the Great Plains and Rocky Mountains from North Dakota and Montana south to Texas, southern Mexico, and Arizona. Flowering is from May to August. Native Americans made a decoction of the roots to treat earaches. Use by wildlife is not reported.

## *Polygonum pensylvanicum* L.

PENNSYLVANIA SMARTWEED, PENNSYLVANIA KNOTWEED, BIGSEED SMARTWEED

**black, brown, green, yellow**

A member of the Polygonaceae (buckwheat family), *Polygonum pensylvanicum* is an annual herb that grows 100–150 cm tall from taproots. Its erect or ascending stems have swollen nodes, are typically reddish or purplish brown when mature, and bear alternate, simple leaves with lanceolate to elliptic blades. The stems are surrounded by an ocrea, as is characteristic of the family. Borne in showy, spicate racemes on elongate peduncles with glandular hairs, the small flowers have only five pink or pinkish white sepals. The petals are missing. In fruit, the shiny black or dark brown achenes are trigonous. Growing in a variety of clayey, loamy, and sandy soils, this species is typically associated with wet sites such as wetlands, borrow ditches, low areas in cultivated fields, and furrows created for tree planting. It is a dominant in the early stages of plant succession. It thrives in disturbed soils and may form large populations that occur across the continent. Flowering is from late June to October. Native Americans made leaf infusions with this plant to treat bleeding, epilepsy, and, when combined with other plants, to aid healing after childbirth. Decoctions were used to treat horses for colic and urinary problems. The achenes are eaten by many species of birds and small mammals. *Polygonum pensylvanicum* is a photosensitizing species and thus may cause skin damage in livestock that eat it. The common name "smartweed" reflects the itching or burning sensation produced by the foliage when handled by sensitive people. Taxonomists differ in their interpretations of the generic limits of *Polygonum*; some divide it into several genera and use the binomial *Persicaria pensylvanica* for this species. Linnaeus spelled the specific epithet *pensylvanica* in favor of the customary *pennsylvanica*.

Flowering plant of *Polygonum pensylvanicum*.

## *Polygonum punctatum* Elliott

DOTTED SMARTWEED, WATER SMARTWEED

**brown, green, yellow**

A member of the Polygonaceae (buckwheat family), *Polygonum punctatum* is a rhizomatous-stoloniferous, perennial herb. Its decumbent or ascending to erect stems are 30–100 cm tall, have swollen nodes, and bear alternate, simple leaves with lanceolate to elliptic blades. The stems are surrounded by an ocrea, as is characteristic of the family. Irregularly spaced along an elongate rachis, the small flowers have only four or five greenish white sepals. The petals are missing. The yellowish punctate glands that dot the surface of the sepals are conspicuous, hence the common name. In fruit, the shiny dark brown achenes are lenticular or slightly trigonous. Typically growing in wet soils or shallow water along streams, at pond margins, or in open swamps, this species occurs across the continent and extends south into South America. Native Americans used its leaves and roots in decoctions to treat stomach pains, swelling of the joints, and other ailments. Its achenes are eaten by wildlife. The common name "smartweed" reflects the itching or burning sensation produced by the foliage when handled by sensitive people. Taxonomists differ in their interpretations of the generic limits of *Polygonum*; some divide it into several genera and use the binomial *Persicaria punctata* for this species.

### ***Polystichum acrostichoides*** (Michx.) Schott

CHRISTMAS FERN

**yellow**

A member of the Dryopteridaceae (wood fern family), *Polystichum acrostichoides* is a perennial, evergreen fern that arises from stout rhizomes. Its large, arching, leathery fronds are 30–80 cm long, linear-lanceolate, and one-pinnately compound. The pinnae are oblong to falcate, with one conspicuous auricle and serrulate-spiny margins. The fertile pinnae are much smaller than the sterile ones and located at the ends of the fronds. The sori

Fronds of *Polystichum acrostichoides*. Photo by Dan Tenaglia.

are confluent and completely cover the abaxial surfaces of the pinnae. Typically growing in rich loamy soils of forest floors, ravines, and shady rocky slopes, this species occurs throughout the eastern half of the continent, reaching southeastern Nebraska and eastern Kansas, Oklahoma, and Texas. It was used extensively by Native Americans, often in combination with other plants. Decoctions, infusions, and poultices of the fronds, rhizomes, and roots were used medicinally both internally and externally to treat a variety of ailments. Ruffed grouse and wild turkeys eat the fronds.

### ***Polytaenia nuttallii*** DC.

PRAIRIE PARSLEY

**brown, red, yellow**

A member of the Apiaceae (carrot family), *Polytaenia nuttallii* is a stout, yellow-green, single-stemmed, perennial herb that arises from a large taproot. Its erect stems are 50–100 cm tall and bear alternate, two- or three-pinnately compound leaves with ovate to oblong leaflets. As is characteristic of the family, the petiole bases partially sheath the stems. Borne in loose compound umbels, the small five-merous flowers have radially symmetrical corollas with free yellow petals and inferior ovaries. In fruit, the schizocarps are elliptic to oblong, flattened, and ribbed. Typically growing as scattered plants or in small populations in dark loamy soils of prairies or open woods, this species is distributed in the center of the continent from Canada to the Gulf of Mexico. Flowering is in May and June. Meskwakis made a decoction of the seeds to treat diarrhea.

Inflorescence of *Polytaenia nuttallii*.
Photo by Bruce W. Hoagland.

### ***Populus deltoides*** Bartram ex Marshall

EASTERN COTTONWOOD

**black, brown, orange, yellow**

A member of the Salicaceae (willow family), *Populus deltoides* is a large deciduous tree 20–30 m tall. It is readily recognized by its massive trunks, 1–2 m in diameter, with deeply furrowed gray bark; its deltoid, crenate leaves with flattened petioles; and its pendulous catkins of capsules dehiscing to release masses of seeds with white hairs (the cotton). Because of a flat place in their petioles, the leaves exhibit a characteristic sidewise flutter in the slightest of breezes, making this tree easy to identify from considerable distances. Often forming gallery forests with distinct cohorts of different heights, this species is characteristically found in the wet soils of sandbars, floodplains, and edges of streams, ponds, and reservoirs. Germination and seedling establishment require wet, barren soils at the time the short-lived seeds are dispersed from the capsules. This species is characteristic of the early stages of plant succession, but plants may persist into the mid stages. The trees are extremely fast-growing, gaining as much as 1–3 m per year. They are also easily propagated by cuttings. Mature trees are fire-tolerant because of their thick bark and can withstand droughty conditions. *Populus deltoides* is distributed throughout the eastern half of the continent as far west as the foothills of the Rocky Mountains. Flowering is in late March and April. Its inner bark contains the glycoside salicin, a precursor of salicylic acid (aspirin), which has been used for millennia to treat headaches, fevers, and inflammation. The soft wood is used for boxes, crates, and pallets and in recent years has become a source of pulp with plantations established in the floodplain of the Mississippi River. Beavers, rabbits, porcupines, and squirrels eat the bark and terminal buds, whereas mule deer, white-tailed deer, elk, and some small mammals consume the twigs and foliage of new growth. The trees are important for turkey roosts and raptor nests, including those of the bald eagle. Limb and trunk cavities also provide excellent den sites for small mammals.

Leaf of *Populus deltoides*.

### ***Potentilla arguta*** Pursh

TALL CINQUEFOIL, PRAIRIE CINQUEFOIL, TALL POTENTILLA

**black, brown, green, orange, red, yellow**

A member of the Rosaceae (rose family), *Potentilla arguta* is a perennial herb that grows 30–100 cm tall from a stout rhizome or branched rootstock. Its erect, viscid-pubescent stems bear alternate, one-pinnately compound leaves with five to eleven elliptic leaflets. The basal leaves are long-petioled, whereas the upper cauline leaves are nearly sessile. Borne in terminal clusters, the five-merous, radially symmetrical flowers have yellowish white to white petals, twenty-five or thirty stamens, and numerous pistils all borne on a saucer-shaped hypanthium. In fruit, the achenes are partially enclosed by the hypanthium. Growing in a variety of soil types in prairies, open woods, and roadsides, this species occurs primarily across the northern half of the continent; it is absent in the southeastern quarter as far west as Texas. Flowering is from late May to July. Native Americans made decoctions and poultices of the roots to treat dysentery, wounds, and other ailments. Use by wildlife has not been reported.

### *Prosopis glandulosa* Torr.

MESQUITE, HONEY MESQUITE, GLANDULAR MESQUITE

**brown, yellow**

A member of the Mimosaceae (mimosa family), *Prosopis glandulosa* is a small, irregularly shaped tree or large shrub that reaches 4–6 m tall. It is readily recognized by its zigzag branches with short spurs and stout spines; its dichotomously forked, compound leaves bearing small linear to oblong leaflets; and its elongate, cylindrical legumes. The yellow-green cast of the foliage and the plant's irregular shape facilitate recognition at a distance. With roots extending downward 30 m or more, this phreatophyte grows in dry sandy, gravelly, or clayey soils. It is characteristic of the mid to late stages of plant succession. Suppression of fire and dispersal of seed via cattle have allowed this tree to invade into prairies, where it is capable of dominating. Long-distance dispersal via cattle may result in disjunct populations. It is distributed in the south-central portion of the United States and adjacent Mexico, and occurs as far north as southern Kansas. Some ecologists now contend that, contrary to popular belief, the geographic range of this species is much the same as it was before the advent of Europeans. It has, however, greatly increased in abundance within its

Tree of *Prosopis glandulosa*.

range. Flowering occurs in late May and June. Considered a troublesome pest, *P. glandulosa* is expensive to control once established; mechanical chaining and aerial spraying with herbicides have been employed. The aromatic wood is used for charcoal and is prized for gunstocks because of the beauty of its grain. Some Native American tribes ground the seeds for flour. White-tailed deer browse the twigs and foliage. Bobwhite quail and mourning doves eat the seeds. The seeds and leaves are also consumed by a large number of small mammals. Cattle browse the taxon's twigs, leaves, and legumes. In older references the binomials *P. juliflora* and *P. chilensis* are used, and the genus is positioned in the subfamily Mimosoideae of the broadly circumscribed family Fabaceae (pea family).

### ***Psoralidium tenuiflorum*** (Pursh) Rydb.

SLIM-FLOWER SCURFPEA, SCURFPEA, WILD ALFALFA, GRAY SCURFPEA, PRAIRIE TURNIP

**brown, green, yellow**

A member of the Fabaceae (pea family), *Psoralidium tenuiflorum* is a single- or multistemmed, perennial herb that grows 40–120 cm tall from a woody rootstock. It is readily recognized by its elongate, scurfy, pubescent stems that are unbranched below and much-branched above; its alternate, one-pinnately compound, gray-green leaves with three or rarely five oblanceolate leaflets; and its terminal racemes of small flowers with light blue, purple, or rarely white, papilionaceous corollas. The black glands of its leaves, bracts, and sepals are conspicuous and diagnostic. The small legumes are one-seeded. Characteristic of the late stages of plant succession, this species occurs as either scattered plants or small populations, in a variety of soil types in prairies and open woods. It is distributed primarily in the center of the continent, from Canada south to Texas, New Mexico, and Arizona and from Illinois and Wisconsin west to Utah and Idaho. Flowering is from June to September. Native Americans prepared infusions, decoctions, and poultices of the stems, leaves, and roots to treat a number of ailments. The leaves were smoked ceremonially and to treat influenza. Eating utensils and sunscreens were also made using this plant. When young, the herbage of *P. tenuiflorum* may be grazed by livestock, but as the plant matures, palatability decreases and its forage value is considered poor. When cured in hay, the foliage is readily eaten. There are unconfirmed reports of toxicity to cattle when consumed in large amounts. Furanocoumarins causing photosensitization are present in other species of the genus. Bobwhite quail and wild turkeys eat the legumes. The binomial *Psoralea tenuiflora* has long been used for this species. Taxonomic work, however, has revealed that *Pso-*

*ralea* is strictly an Old World genus characterized by a cupule below each flower. New World species of scurfpea lack this cupule and are therefore classified in the genera *Psoralidium*, *Pediomelum*, *Orbexilum*, and *Hoita*.

### *Ptilimnium nuttallii* (DC.) Britton

MOCK BISHOP'S-WEED

**brown, green, yellow**

A member of the Apiaceae (carrot family), *Ptilimnium nuttallii* is a slender, glabrous, annual herb that arises from fibrous roots. Its erect, branched stems are 30–60 cm tall and bear alternate, one- or two-pinnately compound leaves with filiform leaflets. As is characteristic of the family, the petioles sheath the stem. Borne in hemispheric compound umbels, the small five-merous flowers have radially symmetrical corollas with free white petals and inferior ovaries. In fruit, the schizocarps are subglobose and conspicuously ribbed. Typically encountered as large populations in the wet soils of low areas in prairies, fields, and barrow ditches, this species occurs in the south-central part of the Midwest south from Illinois, Missouri, and Kansas. Use by Native Americans has not been reported, nor has use by wildlife.

### *Pyrrhopappus grandiflorus* (Nutt.) Nutt.

MORNING STAR, FALSE DANDELION, TUBEROUS FALSE DANDELION, TUBEROUS DESERT-CHICORY

**brown, green, orange, yellow**

A member of the Asteraceae (sunflower family), *Pyrrhopappus grandiflorus* is an acaulescent, perennial herb that arises from a cylindrical rootstock ending in a spherical tuber 4–10 cm below the soil surface. It forms a dense rosette of oblanceolate, undulate-parted or deeply dentate leaves. The flowers are borne in a solitary head that terminates a peduncle 20–30 cm tall. Occasionally the peduncles have a few scattered bracts. Bearing numerous bright yellow ligulate florets, the heads are cylindrical with thirteen to twenty-one phyllaries that reflex in fruit. The achenes are beaked and terminate in a pappus of numerous capillary bristles. This species is often mistaken for *Taraxacum officinale* (dandelion), but the two differ in the appearance of their leaves and the colors of their co-

Head of *Pyrrhopappus grandiflorus*.

rollas and pappus. *Pyrrhopappus grandiflorus* grows in a variety of soil types in prairies and periodically disturbed areas such as rights-of-way, parks, and flower beds. It occurs in the southern Great Plains—Texas, New Mexico, Oklahoma, and southern Kansas—but is increasing in abundance and slowly extending its range northward. Although its specific use by Native Americans has not been reported, other species of the genus have been used as food and medicine. The binomial *P. scaposus* appears in the older literature.

## *Ratibida columnifera* (Nutt.) Wooten & Standl.

MEXICAN HAT, PRAIRIE CONEFLOWER, UPRIGHT PRAIRIE CONEFLOWER

**black, brown, green, yellow**

A member of the Asteraceae (sunflower family), *Ratibida columnifera* is a perennial herb that grows 30–100 cm tall from a vertical woody rootstock. Its erect, branched stems are deeply grooved, strigose-hirsute, and bear alternate, simple, subsessile leaves with pinnatifid blades. Borne at the ends of elongate peduncles, the inflorescences are showy, solitary, cylindrical heads with both ray and disk florets. The three to eleven ray florets droop and are yellow or yellow with a red-brown spot at the bases of the corollas. The numerous disk florets are brown or red-brown. Although most common in the calcareous clay-loamy soils of prairies, this species is also adapted to sandy soils. Characteristic of the early to mid stages of plant succession, it thrives in areas with occasional disturbance such as the mowed margins of highway rights-of-way. It occurs in the Great Plains and tallgrass prairie from Alberta, Saskatchewan, and Manitoba south to northwestern Louisiana, eastern New Mexico, and northern Mexico. Flowering is from June to September. Native Americans used the plant to make teas and to cure headaches, stomach pains, and hemorrhages. Solutions made from boiled stems and leaves were applied to rattlesnake bites and skin affected by poison ivy. This species is a popular garden ornamental. Its forage value for cattle, sheep, and wildlife is fair. The achenes are eaten by wild turkeys. Some taxonomists use the binomial *R. columnaris* for the taxon. The common name "coneflower" is also applied to species of *Echinacea*.

Heads of *Ratibida columnifera*.
Photo by Adam K. Ryburn.

### *Rhus copallinum* L.

WINGED SUMAC, SHINING SUMAC, FLAMELEAF SUMAC, DWARF SUMAC, WING-RIB SUMAC

**black, brown, green, yellow**

A member of the Anacardiaceae (cashew family), *Rhus copallinum* is a shrub or rarely a small tree 1–3 m tall. It is readily recognized by its thicket-forming habit; its milky sap; its brown pubescent branches bearing alternate, one-pinnately compound leaves with seven to seventeen entire leaflets that have winged rachises and that are glossy green above and dull green below; and its dense terminal panicles of bright red pubescent drupes. Its foliage typically turns a brilliant red in autumn. Adapted to a variety of soil types, this species usually occurs in open sites, especially at the interface between forest and prairie. In forests it is a plant of the early stages of plant succession; in prairies it is characteristic of the late stages. Colonizing areas by means of seeds, it establishes extensive thickets via its spreading rootstocks. It sprouts vigorously after fire but declines with annual burning. *Rhus copallinum* occurs throughout the eastern half of the continent from New England to Texas. Flowering is in June, with mature fruits present in August and September. Native Americans used all parts of the plant medicinally and used the drupes for food. Soaking the mature drupes in cold water produces a lemonade-like beverage. Dense thickets of this plant provide cover for small mammals, birds, and white-tailed deer. Its drupes are eaten by a variety of upland game birds, songbirds, and small mammals. Although the fruits, which are high in vitamin A, are not highly preferred by wildlife, they persist on the naked branches well into winter and become increasingly important when other foods are exhausted. Mule deer, white-tailed deer, and rabbits browse the twigs and foliage. The specific epithet is incorrectly spelled *copallina* in many books.

Leaves of *Rhus copallinum*.

## *Rhus glabra* L.

SMOOTH SUMAC, SCARLET SUMAC, DWARF SUMAC, WHITE FLAMELEAF SUMAC

**black, brown, green, red, yellow**

A member of the Anacardiaceae (cashew family), *Rhus glabra* is a shrub or rarely a small tree 1–3 m tall. It is readily recognized by its thicket-forming habit; its milky sap; its tan-gray, glaucous, glabrous branches that bear alternate, one-pinnately compound leaves with eleven to thirty-one leaflets that are conspicuously serrate and bicolor; and its dense, terminal panicles of bright red pubescent drupes. Its foliage typically turns a brilliant red in autumn. Adapted to a variety of soil types, this species typically occurs in open sites, especially at the interface between forest and prairie. In forests it is a plant of the early stages of plant succession; in prairies it is characteristic of the late stages. Colonizing areas by means of seeds, it establishes extensive thickets via its spreading rootstocks. It sprouts vigorously after fire but declines with annual burning. The species is distributed across the continent, with the exception of California, the northwest and northeast corners, and areas along the Gulf of Mexico. Flowering is in May and June, with mature fruits present in August and September. Native Americans used all parts of this plant medicinally and used the drupes for food. Soaking the mature drupes in cold water produces a lemonade-like beverage. Because of its brilliant autumn foliage, *R. glabra* is sometimes cultivated as an ornamental. Dense thickets of it provide cover for small mammals, birds, and white-tailed deer. The drupes are eaten by a variety of upland game birds, songbirds, and small mammals. Although the fruits, which are high in vitamin A, are not highly preferred by wildlife, they persist on the naked branches well into winter and become increasingly important when other foods are exhausted. Mule deer, white-tailed deer, and rabbits browse the twigs and foliage. Following fire, the foliage becomes more palatable and contains increased amounts of crude protein and phosphorous.

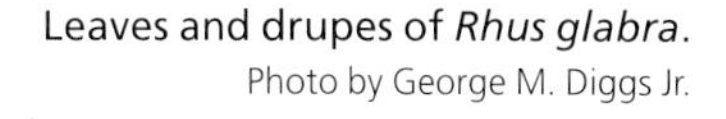

Leaves and drupes of *Rhus glabra*.
Photo by George M. Diggs Jr.

## *Robinia hispida* L.

BRISTLY LOCUST, ROSE ACACIA

**brown, green, orange, yellow**

A member of the Fabaceae (pea family), *Robinia hispida* is a rhizomatous shrub 1–3 m tall, often forming small thickets. Its stems, peduncles, calyces, and legumes are conspicuously hispid, hence the name "bristly locust." The alternate, one-pinnately compound leaves have seven to thirteen ovate-oblong to suborbicular leaflets. Its numerous three- to ten-flowered racemes of large, pink to rose or reddish purple, papilionaceous flowers are conspicuous. Despite the profusion of flowers, few legumes are produced, and those that do occur typically lack seeds. This species is believed to be native to Virginia, the Carolinas, and Georgia but has been widely introduced throughout the eastern two-thirds of the continent as an ornamental and ground cover on mine spoils. In the Midwest it is sometimes encountered at abandoned home sites. Flowering is in June and July. Cherokees used the roots as an emetic and toothache remedy and used the wood to make buildings and implements.

Flowers of *Robinia hispida*.
Photo by George M. Diggs Jr.

## *Robinia pseudoacacia* L.

BLACK LOCUST, POST LOCUST, YELLOW LOCUST, LOCUST-TREE, FALSE ACACIA

**black, brown, green, purple, yellow**

A member of the Fabaceae (pea family), *Robinia pseudoacacia* is a medium tree. Its gray bark is deeply furrowed, and its zigzag branchlets are angled, bearing pairs of spines at the nodes. The alternate, one-pinnately compound leaves have seven to nineteen elliptic to oval leaflets. The numerous pendulous racemes of large, white to cream, papilionaceous flowers are conspicuous, as is the cluster of papery brown legumes that persist on branches in early winter. In forests this species is characteristic of the mid stages of plant succession. It is adapted to a variety of clayey, loamy, and sandy soils. Fast-growing and short-lived, it may form dense groves of spindly trees due to root sprouting. It also sprouts vigorously after cutting or fire. It is thus able to escape from windbreaks or woodlots, invading and

Leaves and flowers of *Robinia pseudoacacia*. Photo by Charles S. Lewallen.

dominating native grasslands. *Robinia pseudoacacia* was distributed originally in the east-central part of the continent from Pennsylvania and Ohio south to Georgia and Alabama and west to Missouri and eastern Oklahoma. Its range is now across the continent because it has been planted extensively in windbreaks, shelterbelts, woodlots, and ornamental gardens. Flowering is in May and June. This species has been used extensively for firewood, fence posts, mine timbers, flooring, and furniture, and has historically been fashioned into pins for telephone pole insulators and treenails for shipbuilding. Native Americans used the bark and leaves for medicine, food, and implements. This species poses a special risk to horses eating its branches and bark, although all animals are probably susceptible. Children eating the bark and seeds have suffered gastroenteric and cardiac dysfunction. The specific epithet is hyphenated as *pseudo-acacia* in older publications.

### ***Rudbeckia grandiflora*** (D. Don) J.F. Gmel ex DC.

ROUGH CONEFLOWER

**black, brown, green, yellow**

A member of the Asteraceae (sunflower family), *Rudbeckia grandiflora* is a perennial herb that arises from a woody rootstock or stout rhizome. Its erect stems are 50–100 cm tall and bear large, alternate, simple leaves with ovate to lanceolate or elliptic blades. The flowers are borne in showy, hemispheric or globose heads that contain both ray and disk florets. Numerous linear, imbricate phyllaries subtend ten to twenty drooping yellow ray florets and numerous purplish or brownish disk florets, each surrounded by a receptacular bract. Adapted to a variety of soil types in open dry woods, this species occurs in an arc from Ohio and Illinois southwest to Louisiana and northeastern Texas. Populations also occur in Georgia. This plant's specific use by Native Americans has not been reported, but other species of the genus were used medicinally.

Head of *Rudbeckia grandiflora*.
Photo by Charles S. Lewallen.

## *Rudbeckia hirta* L.

BLACK-EYED SUSAN, BROWN-EYED SUSAN, HAIRY CONEFLOWER

**black, brown, green, purple, yellow**

A member of the Asteraceae (sunflower family), *Rudbeckia hirta* is a hirsute-hispid, annual, biennial, or short-lived perennial herb that arises from a taproot. Its erect, branched stems are 30–100 cm tall and bear alternate, simple, sessile or subsessile leaves with oblanceolate to elliptic blades. The flowers are borne in showy, hemispheric or globose heads that contain both ray and disk florets. Numerous linear, imbricate phyllaries subtend eight to twenty-one reflexed, yellow or yellow-orange ray florets and numerous brown or red-brown disk florets each surrounded by a receptacular bract. A dominant of the early stages of plant succession, this species is adapted to a variety of clayey, loamy, and sandy soils. It increases in abundance with disturbance, often forming large populations in heavily grazed prairies and pastures, recently logged sites, along roadsides, and

Heads of *Rudbeckia hirta*.
Photo by Robert J. George.

in waste areas. It thrives following periodic fire in open dry woodlands and summer or fall fires in prairies. It occurs across the continent. Flowering is from June to October. Native Americans prepared infusions and decoctions of the roots, stems, leaves, and florets to treat a variety of ailments. Today, *R. hirta* is prized as both a wildflower and a garden ornamental. Bobwhite quail, wild turkeys, and several species of songbirds consume its achenes. In winter the green basal leaves are important forage for white-tailed deer, though limited in abundance. At other times of the year, plants are of moderate preference. They provide poor to fair forage for livestock.

### ***Rumex altissimus*** A.W. Wood

PALE DOCK, SMOOTH DOCK, PEACH-LEAF DOCK

**brown, green, orange, yellow**

A member of the Polygonaceae (buckwheat family), *Rumex altissimus* is a stout, perennial herb up to 100 cm tall. Its erect stems are jointed with swollen nodes, ribbed, and often branched below the inflorescence. The alternate, simple, entire leaves are lanceolate, with long petioles near the stem bases. The inflorescences are loose panicles of verticillate racemes of small, pale green, trigonous flowers with six sepals in two-series—the outer three small and linear, the inner three large and cordate. In fruit, the trigonous achenes are dark brown and remain enclosed by the calyces. Typically encountered in wet soils or in the shallow standing water of swamps, marshes, and borrow ditches, this species occurs across the eastern two-thirds of the United States, reaching Arizona in the southwest and the Dakotas in the northwest. Flowering is April to July. Native Americans used this species medicinally. Specific use by wildlife is not reported, but the achenes are likely eaten by waterfowl, as are those of other species of the genus.

### ***Sabatia campestris*** Nutt.

PRAIRIE ROSE GENTIAN, MEADOW PINK, WESTERN MARSH PINK, TEXAS STAR

**brown, green, orange, yellow**

A member of the Gentianaceae (gentian family), *Sabatia campestris* is a slender, glabrous, annual herb that arises from a taproot. Its erect, branched stems are 10–35 cm tall and bear opposite, simple, sessile leaves that clasp the stem and have ovate to lanceolate, entire blades. Borne singly at the branch ends, the showy flowers have a five-winged calyx tube and radially symmetrical corolla with five fused petals that are roseate to pink with a yellow or greenish spot at their bases. The bright yellow anthers are quite conspicuous, coiling circinately after dehiscence. The stigma lobes also coil about each other after anthesis. Growing in clayey and sandy soils of prai-

ries, forest edges, and roadsides, this species occurs in the lower Midwest from Illinois and Iowa south to Texas, Louisiana, and Mississippi. Populations have been reported in New England as well. Flowering is from April to July. This plant's specific use by Native Americans has not been reported, but Pennsylvanian colonists in the late 1700s used other species of the genus medicinally (Coffey 1993).

Flower of *Sabatia campestris*.
Photo by George M. Diggs Jr., courtesy of the Botanical Research Institute of Texas.

### *Salix caroliniana* Michx.

CAROLINA WILLOW, COASTAL PLAIN WILLOW, LONG-PEDICELLED WILLOW, SOUTHERN WILLOW

**brown, green, orange, red, yellow**

A member of the Salicaceae (willow family), *Salix caroliniana* is a dioecious shrub or small tree up to 9 m tall. Its gray to blackish bark is smooth or furrowed into broad scaly ridges, and its limber twigs are brown with broadly U-shaped leaf scars. The alternate, simple leaves have lanceolate, serrulate blades. They are conspicuously bicolor—green above and whitish below.

Leaves and branches of *Salix caroliniana*. Photo by Bruce W. Hoagland.

The imperfect flowers are borne in terminal, elongate catkins. In fruit, the small capsules contain numerous tiny comose seeds. Growing in wet soils of stream banks, swamps, and sand and gravel bars, this species is commonly encountered throughout the southeastern quarter of the United States extending as far north and west as Pennsylvania, southeastern Nebraska, and central Texas. Flowering is in April and May. As is characteristic of the genus, this species contains the glycoside salicin, a precursor of salicylic acid (aspirin), which has been used for millennia to treat headaches, fevers, and inflammation. Native Americans made decoctions and infusions of the bark, roots, and leaves to treat these and other problems. It was also used as an emetic and for ceremonial purposes, and the wood was used to make toys. A variety of animals eat the buds, bark, leaves, and catkins.

### ***Salix exigua*** Nutt.

SANDBAR WILLOW, COYOTE WILLOW, NARROW-LEAF WILLOW

**black, brown, green, yellow**

A member of the Salicaceae (willow family), *Salix exigua* is a thicket-forming shrub 1–3 m tall. Its gray bark is smooth or somewhat fissured, and its slender twigs are reddish or yellowish brown with broadly U-shaped leaf scars. The alternate, simple leaves have linear, sparsely dentate blades. They are gray-green or yellow-green above and below. The imperfect flowers are borne in terminal, elongate catkins. In fruit, the small capsules contain numerous tiny comose seeds. Growing in wet soils of riverbanks, sandbars, and silt flats, this species has perhaps the greatest geographical range of all willows in North America. It occurs from central Alaska to New York and Louisiana and is a characteristic shrub along streams, especially in the Great Plains and Southwest. Germination and seedling establishment require wet, barren soils at the time the short-lived seeds are dispersed from the capsules. As is characteristic of the genus, *S. exigua* contains the glycoside salicin, a precursor of salicylic acid (aspirin), which has been used for millennia to treat headaches, fevers, and inflam-

Leaves and young branch of *Salix exigua*.

mation. Native Americans made decoctions and infusions of the bark, roots, and leaves to treat these and other ailments. They also used this plant for building materials, bedding, baskets, matting, and cordage. Plants were fed as fodder to their domesticated animals. A variety of animals eat the buds, bark, leaves, and catkins, and use the thickets for cover. Taxonomists now recognize *S. exigua* and *S. interior* as a single species.

### *Salix nigra* Marshall

BLACK WILLOW

**black, brown, green, orange, red, yellow**

A member of the Salicaceae (willow family), *Salix nigra* is a small or medium dioecious tree up to 20 m tall, with a stout, typically leaning and somewhat twisted trunk and irregular crown. Its dark brown bark is deeply furrowed, and its slender, shiny, light brown twigs are marked by broadly U-shaped leaf scars. The alternate, simple leaves have falcate-lanceolate, serrate blades. The imperfect flowers are borne in terminal catkins. In fruit, the small capsules contain numerous tiny comose seeds. *Salix nigra* often forms gallery forests with distinct cohorts of different heights and is characteristic of the early stages of plant succession. It is typically found in the wet soils of sandbars and floodplains and at the edges of streams, ponds, and reservoirs. This species sprouts prolifically after fires, to which it is highly susceptible. Germination and seedling establishment require wet, barren soils at the time that the short-lived seeds are dispersed from the capsules. *Salix nigra* is distributed primarily in the eastern half of the continent, reaching Minnesota and southeastern Nebraska in the northwest and California and Mexico in the southwest. Flowering is in April and May. The inner bark contains the glycoside salicin, a precursor of salicylic acid (aspirin), which has been used for millennia to treat headaches, fevers, and inflammation. The soft wood is

Mature tree of *Salix nigra*.

used for boxes, crates, furniture frames, cabinets, woodenware, pulp, and pallets. A variety of animals eat the buds, bark, leaves, and catkins. This species is also important as nesting cover for wetland birds. It is considered a good honey plant.

## ***Sambucus canadensis*** L.

ELDERBERRY, ELDER, SWEET ELDER

**brown, yellow**

A member of the Caprifoliaceae (honeysuckle family), *Sambucus canadensis* is a shrub 1.5–2.5 m tall that spreads by suckers to eventually form thickets. Its erect, branched stems have glabrous, light brown or grayish brown twigs that are malodorous when bruised. The large drooping leaves are opposite and one-pinnately compound with five to eleven lanceolate or ovate-elliptic leaflets. Borne in large, terminal, compound cymes, the fragrant white flowers are five-merous with rotate corollas, fused petals, and inferior ovaries. Purple-black drupes are produced in abundance. This species is a classic indicator of moist, deep, fertile soils and is found at the edges of streams, seeps, ponds, and ditches. It is occasionally encountered at the margins of moist, open, upland woods. Characteristic of the early stages of plant succession, it is distributed throughout the eastern half of the continent as far west as Manitoba, the Dakotas, Nebraska, Kansas, Oklahoma, Texas, and eastern Mexico. Flowering is in May and June. Native Ameri-

Flowers of *Sambucus canadensis*. Photo by Charles S. Lewallen.

cans used this species extensively. Decoctions, infusions, and poultices were made of various parts of the plant to treat numerous ailments. It was also used to produce food, toys, games, and implements. Today many people relish the jellies and wines made from the drupes. High in crude fat and phosphorus, these drupes are also eaten by upland game birds, songbirds, raccoons, opossums, small mammals, and white-tailed deer. Elderberry thickets are excellent nesting cover for many birds.

### *Sanguinaria canadensis* L.

BLOODROOT, RED PUCCOON

**brown, orange, yellow**

A member of the Papaveraceae (poppy family), *Sanguinaria canadensis* is a rhizomatous, acaulescent, perennial herb with red sap. Its stout rhizome is 6–15 mm in diameter and sends up only a single leaf and one or rarely two large white flowers on an elongate peduncle. The reniform to suborbicular leaf is palmately three- to seven-lobed. Comprising two sepals and eight or sometimes twelve, fourteen, or sixteen petals, the flowers are conspicuous in early spring. This species grows in rich, loamy, often rocky soils of forests and woodlands, occurring across the eastern half of the continent. It is a harbinger of spring, flowering from late February to April, typically before the forest canopy closes. Although *S. canadensis* is typically listed as a toxic plant, extracts from its rhizomes have long been used both in Native American and folk medicines and as an ingredient in over-the-counter oral hygiene products (Mahady, Schilling, and Beecher 1993). Its use is somewhat controversial, and in the popular press, articles either extol its virtues or warn of its toxicity. This plant does contain protoberberines and benzophenanthridines, groups of compounds known to have toxic effects (Preininger 1986).

Flowers of *Sanguinaria canadensis*.
Photo by Paul Buck.

### *Sapindus drummondii* Hook. & Arnold

SOAPBERRY, WESTERN SOAPBERRY, WILD CHINABERRY

**brown, green, orange, yellow**

A member of the Sapindaceae (soapberry family), *Sapindus drummondii* is a small to medium tree 4–12 m tall. It is readily recognized by its gray-green twigs bearing one-pinnately compound leaves with eleven to nineteen falcate leaflets and triangular leaf scars; its large paniculate cymes of small white or greenish white flowers; and its clusters of translucent, amber, one-seeded berries that persist through winter and gradually turn black. Adapted to a variety of soil types, this species may grow as solitary trees or more commonly as dense groves, especially in sand. It is often associated with moist riparian areas but may also occur on upland sites. It is characteristic of the mid stages of plant succession and is distributed in the southwestern portion of the continent from southwestern Missouri and southeastern Colorado to Louisiana, Texas, northern Mexico, and Arizona. Flowering is from May to July. The fruits mature in September and October and persist on leafless branches until the following spring. As the common name "soapberry" indicates, the leaves and berries contain saponins and produce a lather when macerated. Sensitive individuals may develop dermatitis after handling the fruits, which have been used to stupefy fish. The saponins also have the potential to cause irritation of the digestive tract if eaten. Native Americans used the species to make medicines to treat arthritis, rheumatism, and kidney disease, and narrow slats of split wood were used to make baskets. This insect-resistant species is now planted as an ornamental and in shelterbelts. Although opossums and some songbirds, including the American robin, eat the berries, *S. drummondii* has limited food value for wildlife. The trees, however, provide cover for nesting, resting, and escape. Taxonomists differ in their interpretation of whether this species is conspecific with *S. saponaria*, which occurs in the West Indies and Central and South America. Those that unite the two species use the name *S. saponaria* var. *drummondii*. Although *S. drummondii* is often called China-

Leaves and berries of *Sapindus drummondii*.

berry, this common name is most appropriately used for the introduced *Melia azedarach*, a member of the Meliaceae (mahogany family).

### ***Sassafras albidum*** (Nutt.) Nees

SASSAFRAS

**black, brown, green, orange, purple, yellow**

A member of the Lauraceae (laurel family), *Sassafras albidum* is a spicy-aromatic, thicket-forming shrub or tree up to 30 m tall. Its gray-brown bark becomes thick and deeply furrowed with age, and its slender greenish branchlets bear alternate, simple leaves that are elliptic to broadly ovate and that often resemble mittens, with one to three large lobes. In autumn, leaves turn a brilliant yellow, orange, or red. Borne in terminal, racemose clusters, the imperfect yellow or yellow-green flowers appear before the leaves. In fruit, the blue-black drupes produced by the pistillate flowers are borne on red pedicels with enlarged apices. Typically growing in moist soils in openings and at the edges of both upland and bottomland forests, fencerows, and abandoned fields, this species occurs throughout the eastern half of North America as far west as southeastern Kansas and eastern Oklahoma and Texas. Flowering is in April and May. The roots and bark are used to make sassafras tea. Numerous eastern tribes of Native Americans used the species extensively, making infusions, decoctions, and poultices to treat ailments such as rheumatism, colds, burns, sores, measles, diarrhea, and fever. In addition, teas were consumed as a general tonic, blood purifier, emetic, and sedative, and crushed leaves were used to flavor food. Early European explorers and colonists shipped quantities of the root bark to Europe, believing it to be a cure-all (Little 1980).

Leaves, fruits, and pedicels of *Sassafras albidum*. Photo by Paul Buck.

### ***Sedum nuttallianum*** Raf.

YELLOW STONECROP

**brown, yellow**

A member of the Crassulaceae (stonecrop family), *Sedum nuttallianum* is a small annual herb that is succulent and typically yellow-green. Its stems are 4–15 cm tall and bear alternate, simple, subterete leaves with obtuse apices. Its small but showy flowers are four- or five-merous with yellow, radially symmetrical corollas, eight or ten stamens, and four or five pistils. In fruit, the follicles diverge widely. Typically encountered growing in thin rocky soils of rock outcrops or shelves, this species occurs in the south-central part of the United States from southwestern Missouri and central Kansas south to Louisiana and Texas. Flowering is from April to July, the plants sometimes completing their flowering and fruiting in a few weeks before the soil of their habitat dries out completely. This plant's specific use by Native Americans has not been reported, but other species of the genus were used medicinally and for food. A number of *Sedum* species are cultivated as ornamentals.

Flowering plants of *Sedum nuttallianum*. Photo by Paul Buck.

### *Senecio riddellii* Torr. & A. Gray

SAND GROUNDSEL, RIDDELL'S GROUNDSEL

**brown, yellow**

A member of the Asteraceae (sunflower family), *Senecio riddellii* is a glabrous, bright green, perennial herb or subshrub that arises from a woody rootstock. Its numerous branched stems, 30–100 cm tall, arch upward and bear alternate, simple, sessile leaves whose blades are pinnately lobed or dissected into linear-filiform lobes. Numerous cylindrical to campanulate heads are borne at branch ends. Their phyllaries are in two distinct series, and they have twelve to fourteen long yellow ray florets and numerous yellow disk florets. In fruit, the cylindri-

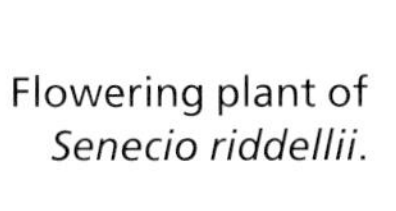

Flowering plant of *Senecio riddellii*.

cal five- to ten-ribbed achenes have a pappus of numerous barbellate hairs. Typically encountered growing in loose sandy soils of dunes and floodplains, this species occurs in the southern Great Plains from South Dakota and Wyoming south to Texas, northern Mexico, New Mexico, and Arizona. Flowering is from August to October. Although this plant's specific use by Native Americans has not been reported, other members of the genus were used medicinally and in ceremonies. *Senecio riddellii* produces compounds collectively known as pyrrolizidine alkaloids, nitrogen-containing ring compounds that are hepatotoxic and cause problems in livestock. Humans also appear to be susceptible, experiencing problems after consuming the plants via herbal teas and dietary supplements (Huxtable 1989).

## ***Silene stellata*** (L.) W.T. Aiton

STARRY CAMPION, WIDOW'S FRILL, WHORLED CAMPION

**yellow**

A member of the Caryophyllaceae (pink family), *Silene stellata* is a perennial herb that grows 30–120 cm tall from a thickened taproot. Its erect, unbranched, puberulent stems are often purple at their bases and bear whorled, simple, sessile or subsessile leaves with lanceolate to ovate-lanceolate, entire blades. This plant's loose, paniculate inflorescence comprises showy, five-merous flowers with a campanulate, inflated calyx and a radially symmetrical corolla of white, fringed petals. In fruit, the open capsules have six apical teeth. Typically encountered growing in rich moist soils of

Flowers of *Silene stellata*. Photo by Charles S. Lewallen.

forest edges and openings, this species occurs throughout the eastern half of the continent, as far west as eastern South Dakota, Nebraska, Kansas, Oklahoma, and Texas. Flowering is from June to August. Meskwakis made poultices of the roots to treat pus-filled swellings. In North Carolina in the 1840s this plant was reputed to be an effective antidote for rattlesnake and copperhead bites (Coffey 1993). Specific use by wildlife is not reported, although seeds of unidentified species of the genus are consumed by various birds.

### ***Silphium integrifolium*** Michx.

PRAIRIE ROSINWEED, WHOLE-LEAF ROSINWEED

**black, brown, green, yellow**

A member of the Asteraceae (sunflower family), *Silphium integrifolium* is a robust, perennial herb that grows 50–150 cm tall from a short rhizome or rootstock. Its erect, typically branched stems bear mostly opposite, sometimes alternate or whorled, simple leaves that are sessile and clasping, with ovate to elliptic, entire or toothed blades. Its inflorescence comprises several large heads on peduncles arising in the upper leaf axils and containing both ray and disk florets subtended by imbricate, ovate to elliptic, ciliate phyllaries. The eleven to thirty-five yellow ray florets are pistillate and produce broadly winged achenes. The numerous light brown or tan disk florets are staminate and subtended by receptacular bracts. Growing in a variety of soil types in prairies and periodically disturbed sites such as roadsides, this species is midwestern in distribution and occurs from the western flanks of the Appalachian Mountains to the western edges of the Great Plains. Flowering is from July to September. Meskwakis used the roots and leaves to treat pain and kidney and bladder problems. As is characteristic of other species of the genus, the large achenes are eaten by various birds.

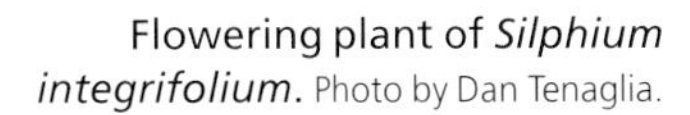

Flowering plant of *Silphium integrifolium*. Photo by Dan Tenaglia.

Flowering plants of *Sisyrinchium angustifolium*.

## ***Sisyrinchium angustifolium*** Mill.

BLUE-EYED GRASS, NARROW-LEAF BLUE-EYED GRASS

**brown, green, yellow**

A member of the Iridaceae (iris family), *Sisyrinchium angustifolium* is a cespitose, bright green, grass-like, perennial herb 10–45 cm tall. Its stems are branched with one or two nodes and bear erect, linear, equitant leaves. Borne in a cluster subtended by a spathe of two conspicuously unequal bracts, the flowers have six pale blue to violet perianth parts, three stamens, and a three-lobed inferior ovary. In fruit, the subglobose capsule is dark brown or black. Growing in a variety of soil types in moist sites of prairies, meadows, and open woods, this species occurs throughout the eastern half of the continent and as far west as central Oklahoma and Texas. Flowering is from April to June. Native Americans made infusions and decoctions of the shoots or roots to treat gastrointestinal problems. Use by wildlife has not been reported.

## ***Smilax bona-nox*** L.

GREENBRIER, CATBRIER, SAW GREENBRIER, BULLBRIER

**brown, yellow**

A member of the Smilacaceae (greenbrier family), *Smilax bona-nox* is a woody vine that arises from a massive knotty rhizome and often forms dense tangles sprawling over other plants or climbing into trees. It is readily recognized by its tough, green, four-sided stems armed with stout prickles; its alternate, simple, net-veined, entire leaves with ovate to cordate blades; and its tendrils. Its small, imperfect, radially symmetrical flowers have six pale green or yellowish green perianth parts. In winter the small clusters of shiny black berries are also diagnostic. Adapted to a variety of soils, this plant typically forms dense thickets in disturbed sites in forests, at forest edges, and in abandoned fields and is characteristic of the mid to late stages of plant succession. When periodically burned or mowed, it may become shrub-like with erect, branched stems. The woodiness of the stems is due to the presence of massive amounts of lignin rather than to the secondary growth produced by a vascular cambium. The rhizomes, or "lignotubers," serve as underground storage organs and can be as large as basketballs, extending a meter below the surface. This species occurs throughout the southeastern quarter of the United States and extends west to southeastern Nebraska and eastern Kansas and Texas. Flowering is from March to June. The berries turn shiny black in fall and typically persist in early winter. Native Americans used this plant for medicine and food. Settlers occasionally broke their plows and harnesses on its massive rhizomes. The tips of the young shoots can be eaten like asparagus. The powdered rootstocks are used to make jellies and to thicken soups, and can be added to water to make a refreshing drink. The berries and herbage are eaten by numerous species of birds and by various mammals, both small and large. This species and other members of the genus are among the most important foods for deer in the southeastern United States. *Smilax bona-nox* also provides important cover for small and medium mammals. *Smilax* was long positioned in the Liliaceae (lily family) but has been segregated into its own family on the basis of its woody tissue, climbing habit, petiolate leaves, and net-veined blades.

Leaf of *Smilax bona-nox*.

## ***Solanum dimidiatum*** Raf.

WESTERN HORSE-NETTLE

**black, brown, green, yellow**

A member of the Solanaceae (nightshade family), *Solanum dimidiatum* is a perennial herb that grows 30–100 tall from deep rootstocks. Its stems and leaves are sparsely covered with stellate hairs and armed with yellow prickles. The alternate, simple leaves have ovate to elliptic-lanceolate blades with five to seven coarse, irregular lobes. Borne in extra-axillary, short-peduncled cymes, the showy flowers have five fused sepals; rotate, radially symmetrical, bluish purple to violet corollas; and five connivent stamens with short filaments and bright yellow anthers opening by apical pores. In fruit, the globose berries are pale yellow and persist on the naked stems in winter. Adapted to clayey, loamy, and sandy soils, this species is characteristic of disturbed soils such as cultivated fields, feedlots, rights-of-way, overgrazed pastures, and waste areas. It may form large populations and occurs throughout the southern half of the continent, most commonly in the west. Flowering occurs from May to October. Although this plant's specific use by Native Americans has not been reported, other species of the genus were used medicinally and as food. Considered a troublesome weed in cultivated fields throughout its range, *S. dimidiatum* is difficult to eradicate because of its deep rootstocks. It produces solanine, a neurotoxic glycoalkaloid that irritates the gastrointestinal tract, and is not normally eaten by livestock.

Flowers of *Solanum dimidiatum*.

### ***Solanum elaeagnifolium*** Cav.

SILVERLEAF NIGHTSHADE, WHITE HORSE-NETTLE

**black, brown, green, yellow**

A member of the Solanaceae (nightshade family), *Solanum elaeagnifolium* is a perennial herb that grows 30–100 cm tall from deep rootstocks. Its erect, branched stems are armed with yellow prickles. It is readily recognized by its alternate, simple, oblong, sinuate-repand leaves; its silvery white, stellate indumentum; its extra-axillary, short-peduncled cymes of blue or violet rotate flowers with bright yellow connivent anthers opening by apical pores; and its yellow berries that persist on naked stems through most of winter. Adapted to clayey, loamy, and sandy soils, this species is characteristic of disturbed sites such as cultivated fields, feedlots, rights-of-way, overgrazed pastures, and waste areas. It may form large populations. Plants occasionally occur in open, disturbed woodlands. *Solanum elaeagnifolium* is distributed in Mexico and the south-central United States from Missouri, Kansas, Arkansas, and Louisiana southwest to Arizona. It is naturalized or adven-

Leaves of *Solanum elaeagnifolium*.

Flowers of *Solanum elaeagnifolium*.
Photo by Robert J. George.

tive in other parts of the continent. Flowering is from May through September. Native Americans used this plant for food, jewelry, and medicine to treat a variety of ailments. Considered a noxious weed of cultivated fields throughout the world, it is difficult to eradicate because of its deep rootstocks. It produces solanine, a neurotoxic glycoalkaloid that irritates the gastrointestinal tract, and is not normally eaten by livestock. However, wood ducks, wild turkeys, bobwhite quail, mourning doves, ring-necked pheasants, raccoons, and other small mammals consume the berries and seeds, and white-tailed deer occasionally browse the foliage.

## ***Solidago missouriensis*** Nutt.

PLAINS GOLDENROD, MISSOURI GOLDENROD

**brown, green, yellow**

A member of the Asteraceae (sunflower family), *Solidago missouriensis* is a perennial herb that grows 30–100 cm tall from rhizomes or branching rootstocks. Its ascending or erect stems bear alternate, simple, thickish, somewhat rigid leaves that are progressively smaller along the stem. The blades are broadly linear to elliptic-lanceolate and entire or serrate. This plant's inflorescence comprises numerous heads borne on one side of recurved branches of a terminal panicle. The small yellow or greenish yellow heads are subcylindrical to campanulate with yellow ray and disk florets subtended by numerous imbricate, linear phyllaries. There are seven to thirteen ray florets and eight to thirteen disk florets. In fruit, the subterete or angled achenes have a pappus of numerous white capillary bristles. Characteristic of the mid to late stages of plant succession, this species is adapted to a variety of soil types and is found in open habitats of prairies and open woods. It increases with disturbance and may form extensive populations in rights-of-way and heavily grazed pastures. It is distributed across the western two-thirds of the continent, with greatest abundance in the Midwest. Flowering is from late July to October. Although specific use of this plant by Native Americans has not been reported, other species of the genus were used to make infusions, decoctions, and poultices to treat a variety of ailments. Because *S. missouriensis* and its relatives flower at the same time as species of *Ambrosia* (ragweeds), they have mistakenly been blamed for causing hay fever in the autumn. In reality they are insect-pollinated. The young foliage is eaten by white-tailed deer, rabbits, prairie chickens, and wild turkeys. Browsing of mature plants has not been reported.

Flowering plants of *Solidago missouriensis*.

## *Solidago rigida* L.

STIFF GOLDENROD, STIFF-LEAF GOLDENROD

**black, brown, green, yellow**

A member of the Asteraceae (sunflower family), *Solidago rigida* is a perennial herb that grows 20–150 cm tall from a branching rootstock. Its erect stems bear alternate, simple, sessile or subsessile leaves with ovate to lanceolate blades. Its large, showy, yellow or greenish yellow heads are borne in a corymbiform inflorescence and have yellow ray and disk florets subtended by numerous imbricate phyllaries. There are seven to four-

Flowering plants of *Solidago rigida*.

teen ray florets and nineteen to thirty-one disk florets. In fruit, the subterete or angled achenes have a pappus of numerous white capillary bristles. Characteristic of the mid to late stages of plant succession, this species is adapted to a variety of soil types and is found in the open habitats of prairies and open woods. It increases with disturbance and may form small populations in rights-of-way and heavily grazed pastures. It is distributed throughout the eastern half of the continent but is more common in the western portion of its range. Flowering is from August to October. Native Americans used this plant to make decoctions, infusions, and lotions to treat gastrointestinal problems, stings, and swellings. Because *S. rigida* and its relatives flower at the same time as species of *Ambrosia* (ragweeds), they have mistakenly been blamed for causing hay fever in the autumn. In reality they are insect-pollinated. Some taxonomists classify stiff goldenrod as *Oligoneuron rigidum*.

### *Sorghastrum nutans* (L.) Nash

INDIANGRASS

**brown, green, yellow**

A member of the Poaceae (grass family), *Sorghastrum nutans* is a perennial herb that grows 90–200 cm tall from short, scaly rhizomes. It is readily recognized by its rhizomatous habit; its blue-green, glaucous foliage; its erect, triangular ligules; and its asymmetrical golden panicles of rames bearing solitary, dorsally compressed, golden brown spikelets with associated ped-

Inflorescence of *Sorghastrum nutans*.

icels. A warm-season species, it is characteristic of the mid to late stages of plant succession and is a dominant grass of the North American tallgrass prairie. It is also found in clearings, glades, and open woods. Occurring in a variety of deep, moist soils, it begins growth in mid spring and flowers profusely in early fall. It responds well to fire. This species is distributed throughout the eastern two-thirds of the continent and south into central Mexico. Use by Native Americans has not been reported. Considered a key management species, *S. nutans* provides high-quality forage when green and is both grazed by livestock and cut for hay. It is a prolific caryopsis producer and a major component of seed mixtures for prairie restoration. White-tailed deer and rabbits consume this plant to a limited extent in spring. Consumption of the caryopses by songbirds and upland game birds has not been reported. Plants provide cover for small mammals. The binomial *S. avenaceum* has been used for this species by some agrostologists.

### *Stillingia sylvatica* Garden ex L.

QUEEN'S DELIGHT

**black, brown, green, orange, red, yellow**

A member of the Euphorbiaceae (spurge family), *Stillingia sylvatica* is a glabrous, monoecious, perennial herb with milky sap that arises from a stout,

Flowering plant of *Stillingia sylvatica*. Photo by Charles S. Lewallen.

woody rootstock. Its numerous ascending or erect stems bear alternate, simple leaves with serrulate to crenate, elliptic to lanceolate or oblanceolate blades. The imperfect flowers lack petals. Small clusters of golden yellow staminate flowers with two stamens are borne on an elongate axis above the pistillate flowers. In fruit, the large three-lobed and three-seeded capsular schizocarps are quite conspicuous. Characteristically encountered in loose sandy soils of dunes, stream banks, and sandy prairies, this species occurs in the southern half of the continent from Virginia to eastern New Mexico and central Mexico. Native Americans made decoctions of the roots to treat a variety of ailments, including diarrhea, menstrual problems, and venereal disease. This species is cyanogenic and may cause irritation of the digestive tract. Its milky sap is irritating to the skin and mucous membranes. Bobwhite quail eat the seeds.

### ***Streptanthus hyacinthoides*** Hook.

SMOOTH TWIST-FLOWER, SMOOTH JEWEL-FLOWER

**yellow**

A member of the Brassicaceae (mustard family), *Streptanthus hyacinthoides* is a glabrous, annual herb 50–100 cm tall. Its erect, generally unbranched stems are often purplish and bear alternate, simple, sessile or short-petiolate leaves with linear-lanceolate, entire or sparsely denticulate blades. Borne in ten- to thirty-flowered racemes, the small magenta to dark purple flowers have an urceolate calyx of four sepals, a radially symmetrical corolla of four free petals, and six stamens (four long and two short). In fruit, the siliques are erect or ascending. Typically growing in sandy soils of prairies or open woods, this species occurs in the south-central United States: Kansas, Oklahoma, Arkansas, Louisiana, and Texas. Use by Native Americans has not been reported, nor has use by wildlife.

Flowering plant of *Streptanthus hyacinthoides*. Photo by Charles S. Lewallen.

## ***Tephrosia virginiana*** (L.) Pers.

GOAT'S RUE, CATGUT, TEPHROSIA, VIRGINIA TEPHROSIA

**green, yellow**

A member of the Fabaceae (pea family), *Tephrosia virginiana* is a perennial herb that grows 20–70 cm tall from a woody rootstock. Its stems and leaves are densely covered with villous to sericeous hairs. The single or clustered stems are erect and bear alternate, one-pinnately compound, sessile or subsessile leaves with fifteen to twenty-five linear-oblong or elliptic leaflets. At the ends of the stems are racemes of large showy flowers with bicolored, papilionaceous corollas. The orbicular, reflexed banners are lemon-yellow or cream, whereas the wings and keel are rose or white. In fruit, the legumes are flattened, straight or curved, gray or tan, and densely villous or sericeous. Occurring as either scattered plants or small localized populations, this species is present in all stages of plant succession and typically occupies stabilized sand dunes and sandy or rocky-loamy soils in open woodlands and prairies. It is somewhat shade-tolerant and persists in developing woodlands. Plants also occur along roadsides and become more abundant following logging or winter fires. This species is distributed throughout the eastern half of the United States and extends west to Iowa, eastern Kansas, Oklahoma, and Texas. Flowering is from May to August, and seeds remain viable in the soil seed bank for many years. Native Americans used the plants, which contain rotenone, as a fish poison and ver-

Flowering plant of *Tephrosia virginiana*.

mifuge. They also made decoctions and infusions of the roots to treat bladder and pulmonary problems and as strengthening tonics. The palatable plants of *T. virginiana* exhibit high nutrient content in spring. When grazed, plants decrease in abundance. Wild turkeys consume the seeds.

## ***Teucrium canadense*** L.

AMERICAN GERMANDER, WOOD SAGE

**black, brown, green, yellow**

A member of the Lamiaceae (mint family), *Teucrium canadense* is a rhizomatous, perennial herb that grows 50–120 cm tall, often forming small colonies. Its solitary, erect stems bear opposite, simple leaves with serrate or crenate, ovate to lanceolate blades. The flowers are borne in a terminal, spicate raceme, and as is characteristic of mints, they have two-lipped corollas. The petals are light rose, lavender, or pale purple. In fruit, four light reddish brown nutlets are present. This species characteristically grows in moist or wet soils of stream banks, marshes, grassy swales, and ditches, occurring across the continent. Flowering is from May to September. Use by Native Americans is not reported, nor is use by wildlife. The species is morphologically quite variable; four varieties are generally recognized.

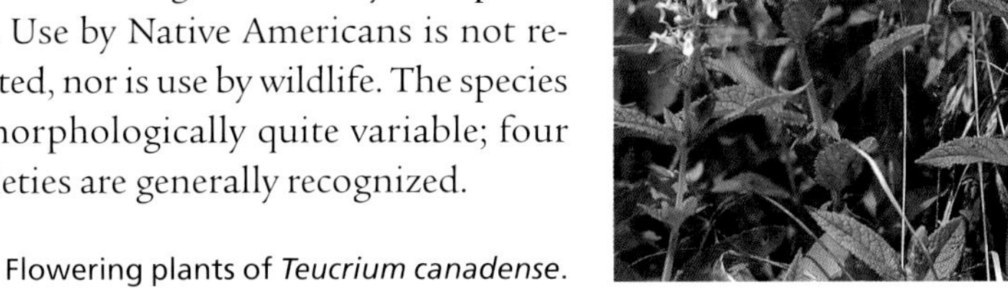

Flowering plants of *Teucrium canadense*.

## ***Thelesperma filifolium*** (Hook.) A. Gray

GREENTHREAD, STIFF GREENTHREAD

**brown, orange, yellow**

A member of the Asteraceae (sunflower family), *Thelesperma filifolium* is an annual or short-lived perennial herb that grows 20–70 cm tall from a taproot or vertical, branching rootstock. Its branched stems bear opposite, one- to three-pinnately dissected leaves with filiform to narrowly linear, entire lobes. Flowers are borne in showy heads at the ends of peduncles that droop when flowers are in bud but that are erect when the heads are open. The urceolate heads have eight yellow or golden yellow ray florets and numerous yellow or reddish brown disk florets. The subtending phyllaries are in two-series—the outer series linear and spreading, the inner series lanceolate to ovate and fused. In fruit, the achenes bear two awns. Adapted to a variety of soil types, this species occurs in periodically disturbed sites of prairies, borrow ditches, and waste areas. It is distributed in the Great Plains from Wyoming and South Dakota south to Texas, Louisiana, and Mississippi. Native Americans made infusions of this plant both as a beverage and as a treatment for tuberculosis. Use by wildlife is not reported.

Head of *Thelesperma filifolium*.
Photo by Robert J. O'Kennon.

### ***Tradescantia occidentalis*** (Britton) Smyth

PRAIRIE SPIDERWORT, PRAIRIE WIDOW'S TEARS
**yellow**

A member of the Commelinaceae (spiderwort family), *Tradescantia occidentalis* is a subsucculent, perennial herb that grows 20–50 cm tall from stout, fleshy roots. Its stems and leaves are glaucous and glabrous. The linear-lanceolate leaf blades are often folded, and the sheaths are inflated. The umbellate clusters of showy, three-merous flowers are borne at the ends of stems and are subtended by a spathe of two leaf-like bracts. The sepals are pubescent, the blue petals broadly ovate. As is characteristic of the genus, the stamen filaments are pilose. In fruit, the capsules contain two to six seeds. Adapted to dry sandy or rocky soils of prairies, plains, and periodically disturbed sites such as borrow ditches and rights-of-way, this species is distributed primarily in the Great Plains, from western Wisconsin and Minnesota south to Louisiana and west to Montana, Utah, and Arizona. Flowering is from March to August. Native Americans made decoctions and infusions of the roots for medicinal use and as stimulatory potions. They also ate the shoots. Specific use by wildlife is not reported. In general, mem-

Flowers of *Tradescantia occidentalis*.

bers of the genus are of little significance to wildlife; the seeds are occasionally eaten by songbirds, and the plants are sporadically browsed by white-tailed deer.

## *Ulmus rubra* Muhl.

SLIPPERY ELM, RED ELM, SOFT ELM

**brown, green, red, yellow**

A member of the Ulmaceae (elm family), *Ulmus rubra* is a large tree up to 20 m tall. It is readily recognized by its open crown of spreading branches and slender scabrous-pubescent twigs; its alternate, simple, ovate to elliptic, scabrous, doubly serrate leaves with acuminate apices and oblique bases; and its suborbicular samaras produced before the leaves in spring. Its winter buds are blunt and densely covered with brownish red hairs. Adapted to a variety of soil types, this species is characteristically encountered in the moist, fertile soils of bottomlands and lower slopes. The leaves and wood decompose to produce rich soil humus. Flowering is from March to May. This species occurs throughout the eastern half of the continent as far west as eastern Saskatchewan, the Dakotas, Nebraska, Kansas, Oklahoma, and Texas. The thick, slightly fragrant inner bark is edible and mucilaginous, hence the common name "slippery elm." Native Americans used this species extensively. Decoctions, infusions, and poultices were prepared to treat a plethora of medical problems. It was also eaten and combined with other foods. The wood and branches were used to make building materials, cordage, tools, matting, and baskets. The wood contains a massive amount of sclerenchyma and is split-resistant. Historically, this wood was used extensively to make railroad ties, hockey sticks, and wagon wheel axils.

Leaf of *Ulmus rubra*.

## *Vaccinium arboreum* Marshall

FARKLEBERRY, SPARKLEBERRY, TREE HUCKLEBERRY, HIGHBUSH HUCKLEBERRY

**black, brown, orange, red, yellow**

A member of the Ericaceae (heath family), *Vaccinium arboreum* is a large shrub or small tree up to 8 m tall, with short, crooked trunks and irregular

Leaves and flowers of *Vaccinium arboreum*. Photo by Dan Tenaglia.

crowns. Its light gray bark is smooth or finely scaly, and its slender, dark red-brown twigs bear alternate, simple, coriaceous leaves with oblanceolate to elliptic, revolute blades. Small, white to rose, urceolate, four- or five-merous flowers are borne in terminal or axillary racemes. In fruit, the black berries dry, harden, and persist on the branches through winter. This is an understory species in upland forests, adapted to sandy and rocky soils. Characteristic of the mid to late stages of plant succession, it is most common on dry, south-facing slopes. In forest openings, plants may form dense stands. Adapted to periodic fire, this species is a fairly prolific sprouter, but it declines if multiple fires occur within three years of each other. It is distributed in the southeastern quarter of the continent as far west as eastern Kansas, Oklahoma, and Texas. Flowering is in May and June. The fruits mature and turn color in August and September. Although *V. arboreum* is related to blueberries and huckleberries, not everyone considers its fruits palatable: some find them tasty, while others consider them dry and seedy. Use by Native Americans is not reported. A variety of songbirds, upland game birds, small mammals, and bears eat the berries.

## ***Vernonia baldwinii*** Torr.

WESTERN IRONWEED, BALDWIN'S IRONWEED

**black, brown, green, yellow**

A member of the Asteraceae (sunflower family), *Vernonia baldwinii* is a perennial herb that grows 60–150 cm tall from a vertical rootstock. It is readily recognized by its erect stems that are unbranched below and branched above; its numerous alternate, simple, ascending, sessile, serrate leaves; and its large inflorescences of numerous heads with purple-lavender disk florets and black-margined phyllaries. In fruit, the cylindrical, ribbed achenes have a double pappus of scales and bristles. The name "ironweed" reflects the tough nature of the stems and the difficulty in pulling this plant from the ground. Adapted to clayey and loamy soils, this species is characteristically present in moist, deep soils of prairies, introduced pastures, and recently logged sites prepared for replanting. It is characteristic of the mid and late stages of plant succession. It increases with heavy grazing and may form rather dense localized populations. Individual plants do not compete well when vigorous stands of grasses are present. *Vernonia baldwinii* is distributed throughout the Midwest from Illinois, Missouri, and Arkansas west to Colorado and south to Texas. Flowering occurs from late July through September. Although this plant's specific use by Native Americans has not been reported, other species of the genus were used to make root decoctions and infusions for treating fevers, chills, menstrual problems, and pain. Because it is not palatable to cattle and other herbivores, the abundance of *V. baldwinii* in a prairie or pasture can be used as an indicator of grazing history. When plants are abundant, they provide limited cover for bobwhite quail and wild turkeys. The small achenes, each with a bristly pappus, are not eaten by wildlife. Butterflies visit the heads.

Flowering plant of *Vernonia baldwinii*. Photo by Wayne J. Elisens.

### *Viburnum rufidulum* Raf.

RUSTY BLACK HAW, SOUTHERN BLACK HAW, NANNY-BERRY

**brown, green, orange, red, yellow**

A member of the Caprifoliaceae (honeysuckle family), *Viburnum rufidulum* is a small tree or large shrub up to 10 m tall, with a short trunk and irregular spreading crown. It is readily recognized by its dark bark; its opposite, lustrous, coriaceous, simple leaves with obovate to elliptic blades; its hemispheric compound cymes of white to cream, five-merous, rotate flowers; and its dark blue or purple-black drupes that persist on leafless branches through early winter. It is also characterized by the dense rusty-brown tomentum of its twigs, the lower surfaces of its blades, and its petioles and pedicels. The leaves typically turn red in fall. Adapted to a variety of soil types and moisture regimes, it occurs in upland and bottomland forest communities as an understory species. It is characteristic of the mid to late stages of plant succession and is also present in forest openings, on stream terraces, and occasionally in prairies. It is distributed in the southeastern quarter of the continent as far north as Ohio, Illinois, and southern Missouri and as far west as southeastern Kansas and eastern Oklahoma and Texas. Flowering is in April and May. The fruits mature and turn color in September. Although specific use of this species by Native Americans has not been reported, other species of the genus were used medicinally and for food. Species of *Viburnum* are prized as ornamentals for their profuse flowering and brilliant fall foliage. The rusty brown tomentum of *V. rufidulum* is quite conspicuous. The sweet drupes are edible and sometimes made into jelly. Songbirds, wild turkeys, bobwhite quail, raccoons, squirrels, and a few other small mammals eat the fruits. White-tailed deer also browse the fruits and exhibit moderate preference for the leaves and twigs in spring, summer, and fall. *Viburnum rufidulum* is morphologically quite similar to *V. prunifolium* (black haw), and the two taxa may be conspecific.

Leaves and flowers of *Viburnum rufidulum*. Photo by Terrence G. Bidwell.

### *Vitis aestivalis* Michx.

PIGEON GRAPE, SUMMER GRAPE

**black, brown, orange, purple, yellow**

A member of the Vitaceae (grape family), *Vitis aestivalis* is a robust, woody vine with high-climbing or sprawling stems that bear branched tendrils opposite the alternate, simple leaves. The young branches are pubescent with whitish or brownish hairs. Leaves are broadly cordate-ovate to suborbicular with three or five shallow or deep lobes. Their surfaces are conspicuously white and pubescent with loose cobwebby hairs. Borne in paniculate cymes, the imperfect or perfect but functionally unisexual flowers are fivemerous and green or greenish yellow. The rachis, branches, and pedicels are tomentose or woolly. In fruit, the black or dark purple berries are 5–14 mm in diameter. This plant is typically found climbing over trees and shrubs and forming large masses of foliage in their crowns. It grows in moist or dry fertile soils of open forests, roadsides, fencerows, and bluffs, occurring throughout the eastern half of the continent as far west as Minnesota, Nebraska, Kansas, Oklahoma, and Texas. Flowering is from May to July. Native Americans made decoctions of the leaves and stems and combined them with other plants to treat a variety of ailments. They also ate the fruits. As with other species of *Vitis*, the berries of *V. aestivalis* are eaten by waterfowl, upland game birds, songbirds, opossums, raccoons, squirrels, foxes, black bears, and white-tailed deer. The foliage and twigs are moderately to highly preferred by white-tailed deer in spring and summer, and moderately preferred in fall. Plants provide nesting cover for squirrels and some songbirds. The fruits make excellent jelly and wine.

Leaves and berries of *Vitis aestivalis*.

# Conversion Tables

## Length

| INCHES | CENTIMETERS |
|---|---|
| 5 | 13 |
| 6 | 15 |
| 7 | 18 |
| 8 | 20 |
| 9 | 23 |
| 10 | 25 |
| 20 | 51 |
| 30 | 76 |
| 40 | 102 |
| 50 | 127 |
| 60 | 152 |
| 70 | 178 |
| 80 | 203 |
| 90 | 229 |
| 100 | 254 |

| FEET | METERS |
|---|---|
| 1 | 0.3 |
| 2 | 0.6 |
| 3 | 0.9 |
| 4 | 1.2 |
| 5 | 1.5 |
| 6 | 1.8 |
| 7 | 2.1 |
| 8 | 2.4 |
| 9 | 2.7 |
| 10 | 3.1 |
| 20 | 6.1 |
| 30 | 9.2 |
| 40 | 12.2 |
| 50 | 15.3 |
| 60 | 18.3 |
| 70 | 21.4 |
| 80 | 24.4 |
| 90 | 27.5 |
| 100 | 30.5 |
| 200 | 61 |
| 300 | 91.5 |
| 400 | 122 |
| 500 | 152.5 |

## Dry Weight

| POUNDS | GRAMS |
|---|---|
| 1 | 454 |

| OUNCES | GRAMS |
|---|---|
| 1 | 28.35 |
| 0.04 | 1 |

## Liquid Volume

| QUARTS | LITERS |
|---|---|
| 1.06 | 1 |
| 1 | 0.95 |

| FLUID OUNCES | MILLILITERS |
|---|---|
| 1 | 29.57 |
| 0.03 | 1 |

# Glossary

**abaxial** (adj.) **1**: with the side of the lateral organ away from the axis; **2**: situated on the underside of the leaf or outside of the petal
**acaulescent** (adj.): stemless or apparently stemless; compare with **caulescent**
**achene** (n.): a small, dry, indehiscent, one-seeded fruit with the ovary wall free from seed except at the funiculus
**acidic** (adj.) **1**: pertaining to habitats or compounds having a pH lower than 7; **2**: having acid properties; compare with **alkaline**
**acuminate** (adj.): tapering to a sharp point with concave sides
**acute** (adj.): tapering to a pointed apex with more or less straight sides
**adjective dye** (n.): a coloring substance that must be applied to a fiber with the additional use of a mordant and that changes in character in relation to the type of mordant used; compare with **mordant dye**
**aerial** (adj.): above the surface of the ground or water
**alkaline** (adj.) **1**: pertaining to habitats or compounds having a pH greater than 7; **2**: having basic properties; compare with **acidic**
**alkaloid** (n.): a basic, nitrogen-containing compound produced by a plant as a secondary metabolite that typically elicits a physiological effect in animals
**allelopathy** (n.): the inhibition of one plant species by chemical compounds produced by another plant species (adj. **allelopathic**)
**alternate** (adj.) **1**: arranged one per node, as pertains to organs such as leaves, branches, flowers, and fruits; **2**: arranged in one floral whorl between organs of an adjacent floral whorl, as when petals alternate with sepals
**alum** (n.): potassium aluminum sulfate, a colorless water-soluble crystal used as a mordant in dyeing
**annual** (adj.): completing its life cycle in one growing season
**anther** (n.): the pollen-bearing portion of a stamen

**anthesis** (n.): the act of flowering (used to designate the flowering period)
**anthocyanin** (n.): a flavonoid chromophore that produces the red, violet, and blue hues observed in most plants
**anthracene** (n.): the chromophore responsible for the durable red dye obtained from madder and cochineal dyes
**apex** (n.): the morphological tip of an organ (pl. **apices**, adj. **apical**)
**appendage** (n.): a secondary part attached to the main structure (adj. **appendicular**)
**areole** (n.): in cacti, a modified axillary bud or short branch bearing one or more spines or glochids
**articulated** (adj.): jointed or appearing to be jointed, with joints sometimes separating at maturity
**ascending** (adj.): rising or extending upward at an oblique angle
**asymmetrical** (adj.) **1**: not divisible into mirror-image halves in any plane; **2**: irregular or unequal in shape or outline; compare with **bilateral** and **radial**
**attenuate** (adj.): tapering gradually to a narrow tip
**auricle** (n.): an ear-like lobe or appendage, typically at the base of an organ (adj. **auriculate**)
**awn** (n.): a bristle-like appendage extending beyond the body of an organ, usually at the tip or from a dorsal surface
**axil** (n.): the upper angle formed between an axis and any organ arising from it, such as that between a leaf and a stem
**axillary** (adj.): borne in the axil of a leaf
**axis** (n.): the principal or central line of development of a plant or organ, such as the main stem (pl. **axes**)
**banner** (n.): in legumes, the adaxial petal in a papilionaceous corolla
**barbed** (adj.): having a point that projects backward, like the barb of a fishhook
**barbellate** (adj.): finely barbed
**bark** (n.): the tissues of a woody plant stem or root that are external to the vascular cambium
**basal** (adj.): attached or situated at the base of a plant or organ
**beak** (n.) **1**: an abruptly narrowed and prolonged tip; **2**: the constricted terminal portion of an ovary, fruit, or style (adj. **beaked**)
**berry** (n.): a fleshy, indehiscent fruit derived from one pistil and having a pulpy or fleshy pericarp (for example, a grape or tomato)
**biennial** (adj.): completing its life cycle in two growing seasons, generally consisting of vegetative growth the first season and reproductive growth the second
**bilabiate** (adj.): two-lipped
**bilateral** (adj.): divisible into mirror-image halves in only one plane; compare with **asymmetrical** and **radial**
**blade** (n.): the thin, expanded portion of a leaf or petal
**bract** (n.): a modified leaf subtending a flower or inflorescence
**branch** (n.): a lateral division of a stem or floral axis
**branchlet** (n.): a small branch, the ultimate division of a branch
**bristle** (n.): a stiff hair
**browse** (vb.): to eat selective parts of a plant, especially young shoots and twigs (n. **browse**)

**bud** (n.): an undeveloped, terminal or axillary organ or shoot with a meristem typically enclosed in bud scales
**bulb** (n.): a short, thick, subterranean stem bearing fleshy leaves (for example, an onion)
**bur** (n.): a fruit or fruiting inflorescence with a rough or prickly covering derived from a pericarp, persistent calyx, or involucre
**bush** (n.): a low, dense shrub profusely branched at ground level
**calyx** (n.): a collective term for sepals (pl. **calyces**)
**campanulate** (adj.): bell-shaped
**canescent** (adj.): gray or white due to short, fine hairs
**canopy** (n.) **1**: a collective term for the crowns of dominant trees or shrubs; **2**: the uppermost, continuous stratum of foliage in a forest, formed by the crowns of trees
**capillary** (adj.): very slender and hair-like
**capitate** (adj.): head-like or forming a head-shaped cluster
**capsule** (n.): a dry, dehiscent fruit derived from a compound pistil and splitting open at maturity along two or more lines of suture
**carotenoid chromophore** (n.): a coloring agent that produces yellow, gold, and orange hues
**caryopsis** (n.): in grasses, a dry, indehiscent, one-seeded fruit with the seed coat fused to the pericarp
**catkin** (n.): a spicate or racemose inflorescence bearing apetalous, unisexual flowers, with the entire inflorescence typically dropping as a unit
**caulescent** (adj.): producing an aerial stem with distinct nodes and internodes; compare with **acaulescent**
**cauline** (adj.): pertaining to or arising from an aerial stem
**cespitose** (adj.): growing in dense clumps or multistemmed tufts
**chlorophyll chromophore** (n.): a coloring agent that produces green hues in plants but that is not easily extracted and affixed to fibers
**chromophore** (n.): an agent found in the objects of nature that influences the light reflection and absorption properties of a surface and thereby determines the perceived color of that surface
**ciliate** (adj.): having a marginal fringe of hairs
**clasping** (adj.): having a foliaceous structure partly or wholly surrounding the stem
**cleft** (adj.): having indentations or incisions that extend about halfway to the middle or the base; compare with **lobed** and **parted**
**clone** (n.) **1**: a population of individuals derived asexually; **2**: one individual of such a population
**clustered** (adj.): positioned closely together, with the specific arrangement not discernable
**cohort** (n.): a group of individuals of the same age
**colony** (n.) **1**: a group of individuals of a species recently established in an area or occupying a particular site; **2**: a group of individuals of a species derived asexually from a parent plant or plants
**color spectrum** (n.): the arrangement of colored light refracted from white light, including red, orange, yellow, green, blue, indigo, and violet
**colorfast** (adj.): having color that is stable and resistant to fading and bleeding

**community** (n.): a ranked category in the classification of vegetation, including all organisms or species occupying a common environment and interacting among themselves, typically characterized by one or more dominant species that create or modify the habitat for the other taxa present

**comose** (adj.): having a tuft of hairs at the end, as with certain seeds

**complementary color** (n.): one of a pair of colors that are located opposite each other on the color wheel and that, when combined, create a grayed hue

**compound** (adj.): divided into or composed of two or more similar parts, as with compound leaves or compound pistils; compare with **simple**

**compressed** (adj.): flattened in one plane

**conduplicate** (adj.): folded lengthwise, usually in two equal appressed halves, with the adaxial surface inside

**connivent** (adj.): converging but not fused

**copper** (n.): blue vitriol (copper sulfate), a mordant used in dyeing

**cordate** (adj.): heart-shaped

**coriaceous** (adj.): leathery or tough but smooth and pliable

**corolla** (n.): a collective term for petals

**corymb** (n.): a short, broad, more or less flat-topped, indeterminate inflorescence in which the outer flowers open first and the lower pedicels are longer (adj. **corymbose**)

**creeping** (adj.): running along the ground and rooting at the nodes

**crenate** (adj.): having a margin with shallow, rounded or blunt teeth

**crown** (n.) **1**: the base of a tufted, herbaceous perennial where the shoot and root systems merge; **2**: the overall appearance or form of the branches of a tree or shrub; **3**: the corona; **4**: in sunflowers, the pappus appearing as a short ring

**cupule** (n.): an involucre composed of hardened bracts that are fused at least at their bases and that partially or completely enclose fruit or fruits

**cyathium** (n.): a modified inflorescence consisting of a pedicelled pistillate flower arising from the center of a cup-like involucre bearing one or more glands on its rim and staminate flowers with one stamen on its inner surface, as is characteristic only of *Euphorbia* and a few related genera (pl. **cyathia**)

**cyme** (n.): a broad, more or less flat-topped, determinate inflorescence with central or terminal flowers opening first (adj. **cymose**)

**deciduous** (adj.) **1**: falling off when no longer functional, as with leaves falling in autumn or floral parts falling after anthesis; **2**: having leaves that drop annually

**decoction** (n.): an extract prepared by placing plant parts, usually roots, bark, twigs, or berries, in cold water, bringing the water to a boil, and then simmering for a length of time; compare with **infusion**

**decumbent** (adj.): lying horizontally on the ground but with the terminal portion ascending

**decurrent** (adj.) **1**: extending downward from the point of insertion; **2**: extending down the side of a style

**deltoid** (adj.): triangular

**dentate** (adj.): having a margin with sharp teeth that point outward rather than forward

**denticulate** (adj.): minutely or finely dentate

**dibromoindigo** (n.): a chromophore used to produce purple dye, found in glandular secretions extracted from certain mollusk species
**dimorphism** (n.): occurrence in two forms
**dioecious** (adj.): having pistillate and staminate flowers borne on different plants; compare with **monoecious**
**direct dye** (n.): a coloring substance that is successfully applied directly to a fiber without treatment with mordants
**disk** (n.) **1**: the more or less fleshy or raised outgrowth of a receptacle; **2**: the flattened adhesive tip of a tendril; **3**: in sunflowers, the central portion of a receptacle bearing disk florets
**dissected** (adj.) **1**: cut in any way; **2**: deeply and/or irregularly cut
**distal** (adj.): situated in the region farthest away from the point of origin or attachment, or the center
**dominant** (n.): an organism or species exerting influence upon a community because of its size, abundance, aerial coverage, or other features
**dorsally compressed** (adj.): having spikelets flattened along the midnerves (backs) of glumes and lemmas
**drupe** (n.): a fleshy, one-seeded, indehiscent fruit with its seed enclosed in a hard or papery endocarp called a pit or stone (for example, a peach)
**dyebath** (n.): a liquid dye solution into which fibers or fabrics are submerged
**ellipsoid** (n.): a solid body that is elliptic in longitudinal section and circular in cross section (adj. **ellipsoidal**)
**elliptic** (adj.): ellipse-shaped
**elongate** (adj.): lengthened, stretched, extended
**emarginate** (adj.): having an apex with a shallow notch
**entire** (adj.): having a continuous margin, without teeth or divisions
**equitant** (adj.): overlapping lengthwise in two ranks, as with the leaves of irises
**erect** (adj.): upright or perpendicular to the surface
**evergreen** (adj.) **1**: having foliage that remains green for more than one growing season and that does not fall en masse; **2**: never lacking green leaves; **3**: remaining green throughout winter
**exfoliate** (vb.): to come off in thin layers, shreds, or plates, as with the flaking bark of sycamores
**exserted** (adj.): extending beyond or protruding from surrounding parts
**falcate** (adj.): sickle-shaped
**fallow** (n.): cropland that is allowed to lie idle for one or more growing seasons
**fascicle** (n.): a tight cluster or bundle of organs appearing to arise from a common point (adj. **fascicled**)
**fertile** (adj.): capable of producing viable seeds
**fibrous** (adj.) **1**: having numerous woody fibers, as with the mesocarp of a coconut; **2**: having roots and their branches that are approximately the same diameter, as with grasses
**filament** (n.) **1**: the part of the stamen that supports the anther; **2**: a thread-like structure (adj. **filamentous**)
**flavonoid chromophore** (n.): a coloring agent that produces yellow, gold, and orange hues
**fleshy** (adj.): firm and juicy

**flexuous** (adj.): having an axis with curves or bends
**floret** (n.) **1**: in grasses, a subunit of a grass spikelet consisting of a flower and two associated bracts, the lemma and palea; **2**: in sunflowers, one flower of a composite head with five fused petals, a highly modified calyx, five stamens fused by their anthers, and an inferior ovary, with florets categorized into three recognized types: disk, ray, and ligulate
**flower** (n.): the reproductive organ of a flowering plant and the site of sporogenesis, gametogenesis, and seed formation, consisting of a determinate shoot axis bearing sepals, petals, stamens, and carpels in spirals or whorls
**follicle** (n.): a dry, dehiscent fruit derived from a simple pistil and dehiscing along one suture line
**forage** (n.): a plant material consumed by herbivores
**forest** (n.): a vegetation type dominated by trees whose crowns typically form a continuous or almost continuous canopy; compare with **woodland**
**free** (adj.): separate, not fused or attached to another organ
**fringed** (adj.): having marginal hairs, bristles, or hair-like structures
**frond** (n.): a fern leaf
**fruit** (n.): a mature ovary and any other structures maturing with it after fertilization
**fugitive** (adj.): lacking fastness and prone to bleeding or fading
**funnelform** (adj.): funnel-shaped, with the corolla tube widening gradually upward toward the apex, as with many morning-glory flowers
**furrow** (n.): a longitudinal channel or groove
**fused** (adj.): joined together
**gibbous** (adj.) **1**: swollen on one side; **2**: having a pouch-like enlargement of the base of an organ
**glabrate** (adj.) **1**: nearly glabrous; **2**: becoming glabrous with age
**glabrous** (adj.): smooth, devoid of pubescence or hairs of any form
**gland** (n.) **1**: an appendage, protuberance, or depression on the surface of an organ or at the end of a hair that secretes usually sticky fluid; **2**: a protuberance that may not be secretory, as with the warty swelling at the base of a leaf blade in *Prunus* species (adj. **glandular**)
**glaucous** (adj.): covered with a fine waxy substance that may be rubbed off and that imparts a whitish or bluish cast to the surface, as with the bloom of a grape or plum
**globose** (adj.): spherical or nearly spherical
**glomerule** (n.): a compact, cymose cluster of flowers or heads subtended by an involucre
**glutinous** (adj.): covered with sticky exudation
**grassland** (n.): a vegetation type dominated by perennial grasses, with few or no trees or shrubs
**graze** (vb.) **1**: to eat almost the entire aerial portion of a plant; **2**: to eat a plant close to the ground
**gynostegium** (n.): in milkweeds, a structure that is formed by the fusion of stamens and style and that is specialized for the mass transfer of pollen by insects
**habit** (n.): the general appearance of a plant (may refer to a single characteristic or to characteristics as a whole)

**habitat** (n.): an environment normally occupied by an individual or species and characterized by a set of physical, chemical, and biotic features

**hastate** (adj.): arrowhead-shaped, with two basal lobes turned outward

**head** (n.) **1**: a dense, spherical, hemispheric, or flat-topped inflorescence of sessile or subsessile flowers; **2**: in sunflowers, a dense, spherical, hemispheric, or flat-topped inflorescence of sessile flowers borne on a common receptacle and subtended by an involucre

**hepatotoxic** (adj.): having toxic effects on the liver

**herb** (n.): a plant with nonwoody aerial stems that die at the end of the growing season

**herbaceous** (adj.): **1**: relating to an herb; **2**: having the texture or color of a leaf; **3**: lacking woody tissue

**herbage** (n.) **1**: a collective term for the stems and leaves of a plant; **2**: the total plant biomass available to an herbivore

**herbivore** (n.): an animal that feeds on plants (adj. **herbivorous**)

**hirsute** (adj.): having long, stiff, coarse, erect or ascending, straight hairs

**hispid** (adj.): bristly and rough, with rigid, long, tapered, erect or ascending, straight hairs

**hispidulous** (adj.): minutely hispid

**hue** (n.): a color category, such as red, blue, or yellow

**husk** (n.): a covering that encloses the fruit, generally derived from a perianth or involucre

**hyaline** (adj.): thin and translucent or transparent

**hypanthium** (n.): the cup-shaped to elongate tubular extension of a floral axis that surrounds the gynoecium, produced by the fusion of the basal parts of the perianth and androecium or occasionally as an expansion of the receptacle (pl. **hypanthia**, adj. **hypanthial**)

**imbricate** (adj.): partially overlapping, as shingles on a roof

**imperfect** (adj.): lacking stamens, pistil, or both

**incised** (adj.): cut sharply, deeply, and usually irregularly

**indehiscent** (adj.): not splitting open to release contents when mature

**indigotin** (n.): a chromophore that is found in species of indigo and that produces a durable blue dye

**indumentum** (n.): a surface covering of hairs, scales, or scurf (adj. **indumented**)

**inferior** (adj.) **1**: having one organ lower than another; **2**: having the ovary below the apparent point of attachment of the other floral organs

**inflorescence** (n.): the arrangement of flowers and accessory parts on an axis

**infusion** (n.): an extract prepared by pouring hot water over plant parts, usually leaves and flowers, and allowing them to sit for a period of time; compare with **decoction**

**involucre** (n.) **1**: a collective term for a dense cluster of bracts subtending a flower or inflorescence; **2**: a collective term for a bract, pair of bracts, or whorl of bracts subtending a flower or inflorescence; **3**: in walnuts, the fused bracts forming a husk enclosing a nut; **4**: in oaks, the fused bracts forming a cap partially enclosing a nut

**iron** (n.): copperas (ferrous sulfate), a metallic salt used as a mordant in dyeing

**jointed** (adj.): having distinct nodes or points of actual or apparent separation

**keel** (n.) **1**: a prominent longitudinal ridge, analogous to the keel of a boat; **2**: in legumes, the two fused abaxial petals in the papilionaceous corolla
**lanceolate** (adj.): lance-shaped
**lateral** (adj.): on or at the side of an axis or organ
**leaf** (n.): a lateral appendage arising from a stem at a node and typically subtending an axillary bud with the primary functions of photosynthesis and transpiration, comprising a blade, petiole, and stipules
**leaflet** (n.): a single segment of a compound leaf
**legume** (n.): a dry, one- or rarely two-locular, two-valved fruit derived from a simple pistil and typically dehiscent along two sutures, as is characteristic of legumes, mimosas, and caesalpinias
**lenticular** (adj.): lens-shaped
**lightfast** (adj.): stable and resistant to fading in the presence of light
**ligule** (n.) **1**: a tongue- or strap-shaped organ; **2**: in sunflowers, a strap-shaped limb of a ray or ligulate floret; **3**: in grasses, a membranous or hairy appendage on the adaxial surface of a leaf at a junction of sheath and blade (adj. **ligulate**)
**limb** (n.): the broadened or flattened part of an organ extending from a narrower base, such as the expanded portion of a fused corolla
**linear** (adj.): long and narrow with margins parallel or nearly so
**lip** (n.) **1**: one of the two projections or segments of a bilabiate corolla or calyx; **2**: a petal modified or differentiated from the others, as with the labellum of an orchid flower
**lobe** (n.): a segment or portion of an organ separated from an adjacent segment by sinuses or clefts (adj. **lobate**)
**lobed** (adj.) **1**: having lobes; **2**: having indentations or incisions that extend less than halfway to the middle or the base; compare with **cleft** and **parted**
**loment** (n.): a flat legume constricted between seeds and disarticulating at constrictions into one-seeded joints when mature
**longitudinal** (adj.): extending along the long axis of an organ
**malodorous** (adj.): having a disagreeable odor
**margin** (n.): the edge or border of an organ
**-merous** (adj.): having a certain number of parts in each whorl of a flower (for example, a three-merous flower has sepals, petals, stamens, and carpels in threes or multiples of three)
**monoecious** (adj.): having pistillate and staminate flowers borne on the same plant; compare with **dioecious**
**mordant** (n.): a chemical substance used in dyeing that combines with both the dyestuff and the material being dyed to deposit insoluble color onto the fiber
**mordant dye** (n.): a coloring substance that must be applied with a mordant and that changes in character with the type of mordant used; compare with **adjective dye**
**multiple** (adj.): produced by ripened ovaries of several to many flowers, as with mulberries
**naked** (adj.): devoid of attached structures, hairs, appendages, or other coverings
**native** (adj.): occurring naturally in an area
**nerve** (n.): a prominent longitudinal vein or slender rib
**net-veined** (adj.): forming a network of anastomosing veins

**neurotoxic** (adj.): causing toxic effects on the nervous system
**neutral** (adj.): lacking functional stamens and pistils
**nitrogen fixation** (n.): a process in which certain free-living and symbiotic bacteria incorporate atmospheric nitrogen into nitrogen compounds
**node** (n.): a point on a stem where leaves, branches, or flowers originate
**nut** (n.): a dry, indehiscent, usually one-loculed, one-seeded fruit derived from a compound ovary with a bony, woody, leathery, or papery pericarp and partially or wholly enclosed in an involucre
**nutlet** (n.) **1**: a dry, indehiscent, one-seeded fruit derived from half of a carpel, as is characteristic of mints, borages, and vervains; **2**: a small nut
**oblique** (adj.): slanted with unequal sides
**oblong** (adj.): rectangular with rounded ends
**obovate** (adj.): egg-shaped with the narrower end basal
**obtuse** (adj.): blunt or rounded at the end with the sides forming an angle of more than 90 degrees
**ocrea** (n.): in knotweeds, the sheath around the stem formed by the fusion of two stipules (adj. **ocreate**)
**opposite** (adj.) **1**: arranged two per node; **2**: having the organs in one floral whorl superposed with those of an adjacent floral whorl, as when stamens are opposite the petals
**orbicular** (adj.): essentially circular in outline, in reference to flat organs
**oval** (adj.): broadly elliptic, the width more than half the length
**ovary** (n.): the basal portion of a pistil containing the ovules and maturing into fruit after fertilization
**ovate** (adj.): egg-shaped in two dimensions and attached at the broader end, in reference to flat surfaces
**ovoid** (adj.): egg-shaped in three dimensions, in reference to solids
**palmate** (adj.): having two or more lobes or segments radiating from the same basal point, like fingers spreading from the palm of a hand
**panicle** (n.): an indeterminate, branched inflorescence comprising a rachis, one branch or a series of branches, and pedicels bearing flowers (adj. **paniculate**)
**papilionaceous** (adj.): having a butterfly-like corolla comprising five petals differentiated into a standard (banner), wings, and keel, as with legumes
**pappus** (n.): in sunflowers, a highly modified calyx of florets, usually appearing in the form of capillary bristles, plumose bristles, scales, awns, or a short crown
**parted** (adj.): having indentations or incisions that extend more than halfway to the middle or the base; compare with **cleft** and **lobed**
**pedicel** (n.) **1**: a stalk bearing an individual flower; **2**: in grasses, a stalk bearing a spikelet (adj. **pedicellate**)
**peduncle** (n.): a primary axis or stalk terminating in an inflorescence (adj. **pedunculate**)
**pendulous** (adj.): hanging or drooping downward
**perennial** (adj.) **1**: completing its life cycle in three or more growing seasons; **2**: continuing to live from year to year (vb. **perennate**)
**perfect** (adj.): having both stamens and pistils
**perianth** (n.): a collective term for the calyx and corolla (if both are present, the perianth is said to be in two-series; if the calyx or corolla are absent, the perianth is said to be in one-series)

**persistent** (adj.): remaining attached
**petal** (n.): one member of the second whorl of floral organs, typically colored and showy
**petaloid** (adj.): petal-like in color, shape, or texture
**petiole** (n.): the stalk of a leaf connecting a blade and stem node (adj. **petiolate**)
**phreatophyte** (n.): a plant with a root system typically in soil saturated with water
**phyllary** (n.): in sunflowers, the involucral bract of a head subtending florets
**pilose** (adj.): having long, soft, erect or ascending, generally straight hairs
**pinna** (n.) **1**: one of the primary divisions of a pinnately compound leaf; **2**: the primary division of a fern frond (pl. **pinnae**)
**pinnate** (adj.): having leaflets borne in two rows along an axis (if one-pinnately compound, leaflets are attached to the rachis; if two- or three-pinnately compound, leaflets are attached to the secondary or tertiary axes)
**pinnatifid** (adj.): pinnately cleft to parted into narrow lobes almost to the midrib
**pistil** (n.): the innermost floral organ, normally differentiated into ovary, style, and stigma, comprising one carpel when simple, two or more carpels when compound
**pistillate** (adj.): bearing a pistil or pistils but lacking functional stamens
**pith** (n.): the centermost tissue of a stem, usually soft or spongy
**plumose** (adj.): indumented in a manner simulating a feather or plume
**pollen** (n.): the mature microspores or developing microgametophytes in seed plants
**pollination** (n.) **1**: in flowering plants, the transfer of pollen from an anther to the stigma of a pistil; **2**: in gymnosperms, the transfer of pollen from a pollen-producing cone directly to an ovule
**polygamous** (adj.): bearing perfect and imperfect flowers on the same plant
**population** (n.): individuals of a species that occupy an area at the same time and that are typically isolated to some degree from other groups
**pore** (n.): an orifice or opening (adj. **poricidal**)
**poultice** (n.): softened plant material applied to a sore or inflamed part of the body, usually heated and spread on a cloth
**prickle** (n.): a small, spine-like outgrowth of bark or epidermis
**primary color** (n.): a color that cannot be produced by combining other colors (red, yellow, and blue are all primary colors)
**prostrate** (adj.): laying flat on the ground
**puberulent** (adj.): minutely pubescent
**pubescent** (adj.) **1**: exhibiting hairs of some kind; **2**: covered with short, soft, erect or ascending hairs
**punctate** (adj.) **1**: having pits, **2**: having colored or translucent, sunken glands
**pyramidal** (adj.): pyramid-shaped
**raceme** (n.): an indeterminate inflorescence consisting of a rachis bearing a number of pedicelled flowers (adj. **racemose**)
**rachis** (n.): the central axis of a compound leaf or compound inflorescence, bearing leaflets or flowers (pl. **rachises**)
**radial** (adj.) **1**: denoting a radius; **2**: developing uniformly on all sides; **3**: divisible into mirror-image halves in two or more planes; compare with **asymmetrical** and **bilateral**

**rame** (n.): an inflorescence consisting of dorsally compressed paired spikelets, one sessile and one pedicelled, borne on a disarticulating rachis, as is characteristic of certain grasses

**rank** (n.): a vertical row along an axis

**ray** (n.) **1**: in sunflowers, a pistillate or neutral floret with a strap-shaped corolla; **2**: one of the radiating stalks arising from the apex of a peduncle of a compound umbel

**receptacular bract** (n.): in sunflowers, a membranous or hyaline scale or bristle arising from a receptacle and subtending the disk floret

**recurved** (adj.): gradually bent or turned backward

**reduction** (n.): the process of reducing a dye to a soluble form for application to a fiber, after which the dye is oxidized to the original insoluble pigment

**reflexed** (adj.): abruptly bent downward or backward

**reniform** (adj.): kidney-shaped

**resin** (n.): a yellowish to dark, sticky, organic exudate (adj. **resinous**)

**reticulate** (adj.): net-veined

**retrorse** (adj.): directed backward or downward

**revolute** (adj.): rolled downward from the edges so that the abaxial surface is partially concealed

**rhizome** (n.): a horizontal underground stem producing aerial stems and leaves along its length or at the end, with or without scale leaves (adj. **rhizomatous**)

**rhomboidal** (adj.): diamond-shaped

**ribbed** (adj.): having prominent ribs or veins

**ridge** (n.): a long, relatively narrow, raised strip with sloping sides

**root** (n.): the descending axis of a plant that grows opposite the stem, without nodes and leaves, typically developing underground and having the primary functions of anchorage and absorption of water and nutrients

**rootstock** (n.): a woody underground stem base or root apex giving rise to aerial growth each growing season

**rosette** (n.): a cluster of leaves radiating from a common center or crown, usually at or near ground level or rarely at the stem apex

**rotate** (adj.): having a circular corolla with a short tube and wide limbs spreading at right angles to the tube

**rugose** (adj.): irregularly and coarsely wrinkled

**saline** (adj.): rich in soluble salts

**salverform** (adj.): having a trumpet-shaped corolla with a slender tube and an abruptly expanded limb at right angles to the tube, as is characteristic of phlox

**samara** (n.): a winged, dry, indehiscent, one-seeded, achene-like fruit, as is characteristic of elms and ashes

**sap** (n.): the aqueous fluid circulated throughout a plant

**saturation** (n.): the intensity, purity, or brightness of a color

**scabrous** (adj.): rough to the touch due to epidermal projections or short stiff hairs

**scale** (n.) **1**: a small, thin, usually scarious leaf or bract; **2**: a small leaf modified for the protection of the bud meristem; **3**: a small, scarious or coriaceous appendage of the tissue within a perianth; **4**: in sedges, a small bract subtending

an individual flower in a spike; **5**: a thin, flat structure of epidermal origin (adj. **scaly**)
**scar** (n.) **1**: a mark or indentation left on a stem by the separation of a leaf or bud scale; **2**: a mark or indentation left on a seed when it detaches
**scarious** (adj.): thin, dry, membranous, semitranslucent, and not green
**schizocarp** (n.): a dry, dehiscent fruit derived from a compound pistil that splits into separate, one-seeded, one-carpellate segments at maturity, as occurs with carrots and geraniums
**scour** (vb.): to clean wool of dirt and oil by washing or treating it with solvents
**scurfy** (adj.): covered with small bran-like scales
**secondary color** (n.): a color produced by combining two primary colors (orange, green, and purple are all secondary colors)
**seed** (n.): a multicellular structure developing from an ovule after fertilization, typically comprising an embryo, seed coat, and storage tissue such as endosperm
**sepal** (n.): one member of the outermost whorl of floral organs, typically green and sometimes leaf-like or petaloid
**sere** (n.) **1**: a succession of plant communities in a given habitat leading to a particular climax association; **2**: one stage in a succession of communities
**sericeous** (adj.): silky, with long, soft, slender, generally appressed hairs
**sericin** (n.): a natural gummy coating on raw silk filaments
**series** (n.): the number of structures standing in an order and related in some fashion (for example, a calyx and corolla are each a series of floral parts)
**serrate** (adj.): having a margin with sharp teeth that point forward
**serrulate** (adj.): minutely serrate
**sessile** (adj.): attached directly at the base, as a leaf without a petiole
**setaceous** (adj.) **1**: bristle-like; **2**: bearing bristles
**sheath** (n.): a long, more or less tubular structure partly or wholly surrounding another organ or part
**shoot** (n.) **1**: a collective term for the stem and leaves; **2**: a young stem or branch arising from roots or older stems or branches
**shrub** (n.): a woody plant with multiple stems at the base and no single trunk
**shrubland** (n.): a vegetation type dominated by shrubs or small shrubby trees whose crowns do not touch
**silicle** (n.): in mustards, a dry, dehiscent, two-carpellate fruit, less than two times longer than wide, with two valves separating from a persistent septum and placentae
**silique** (n.): in mustards, a dry, dehiscent, two-carpellate fruit, more than two times longer than wide, with two valves separating from a persistent septum and placentae
**simple** (adj.): not divided into distinct parts or segments, as with a leaf with an undivided blade or a pistil with one carpel; compare with **compound**
**sinuate** (adj.): having a wavy margin in a horizontal plane that is alternatively concave and convex; compare with **undulate**
**skein** (n.): a continuous strand of yarn arranged in a coil
**solitary** (adj.): single
**sorus** (n.): a cluster of sporangia in ferns (pl. **sori**)

**spathe** (n.): a bract or pair of bracts surrounding or subtending a solitary flower, spadix, or other inflorescence

**spathulate** (adj.): spatula-shaped

**spicate** (adj.): spike-like

**spike** (n.): an inflorescence consisting of a central rachis bearing one or more sessile flowers

**spikelet** (n.) **1**: in grasses, a highly modified inflorescence typically consisting of two glumes, one or more florets, and rachilla; **2**: in sedges, a highly condensed spike consisting of flowers and subtending bracts borne on a short axis

**spine** (n.) **1**: a modified leaf or stipule appearing as a sharp, stiff projection from a stem; **2**: a structure with the appearance of a spine (adj. **spinose**)

**spreading** (adj.): extending more or less horizontally outward

**sprout** (n.): an aerial stem arising from roots

**spur** (n.) **1**: a short, compact branch with little or no internodal development, often bearing fascicles of leaves or flowers; **2**: a tubular or sac-like basal extension of a sepal or petal (adj. **spurred**)

**stamen** (n.): a floral organ that produces pollen, normally consisting of a filament and anther (adj. **staminal**)

**staminate** (adj.): bearing stamens but lacking a functional pistil

**stand** (n.) **1**: a unit of vegetation comprising a single species; **2**: a unit of vegetation that, together with other units having similar combinations of species, forms an association

**stellate** (adj.) **1**: star-shaped; **2**: having hairs with several branches that radiate from one point

**stem** (n.): the ascending, typically aboveground axis of a plant that grows opposite the roots and bears and supports leaves, flowers, and fruits

**sterile** (adj.): not producing viable seeds

**stigma** (n.): the portion of a pistil that receives pollen, typically situated at the end of the style (adj. **stigmatic**)

**stipe** (n.) **1**: the stalk of a pistil or other organ or gland; **2**: the petiole of a fern frond (adj. **stipitate**)

**stipule** (n.): one of a pair of appendages borne at the base of a leaf (adj. **stipulate**)

**stoloniferous** (adj.): producing elongate, horizontal, aboveground stems (stolons), which typically root at nodes and give rise to new shoots

**striate** (adj.): having narrow, usually parallel lines or grooves

**strigose** (adj.): having long, stiff, sharp, straight or curved, appressed hairs

**style** (n.): a slender stalk connecting the stigma to the ovary (adj. **stylar**)

**substantive dye** (n.): a coloring substance that is successfully applied directly to a fiber without treatment with mordants

**subtend** (vb.): to enclose or extend beyond a structure in its axil, as with a bract below a flower or a leaf below a bud

**succession** (n.) **1**: the gradual and predictable progression of changes in the composition of a community that occurs during the development of vegetation in the area; **2**: progressive changes in vegetation from the initial colonization to the climax typical of the area; **3**: the process of continuous colonization and the extinction of species at a particular site; **4**: the chronological distribution of organisms or species in an area

**succulent** (adj.): firm and juicy
**suffrutescent** (adj.): obscurely or somewhat shrubby, slightly woody at the base, with distal herbaceous portions dying back to the woody base
**suture** (n.): a junction or seam of fusion, a line of opening or dehiscence on a fruit or anther
**syncarp** (n.): a compound fruit derived from the fused pistils of two or more flowers (for example, a mulberry)
**tannic acid** (n.): a naturally occurring mordant found in the bark and fruit of many plants, especially oaks and sumacs
**taproot** (n.): a persistent, well-developed primary root, generally larger than the secondary roots that arise from it
**taxon** (n.): a general taxonomic term for any group of any rank (pl. **taxa**)
**tendril** (n.): a rotating or twisting thread-like extension by which a plant grasps an object for support, which morphologically may be a modified stem, leaf, leaflet, or stipule
**terete** (adj.) **1**: cylindrical; **2**: circular in cross section
**terminal** (adj.): situated at the tip or apex of a stem or summit of an axis, as with the terminal leaflet of a one-pinnately compound leaf
**tertiary color** (n.): a color produced by combining a primary color with one of its adjacent secondary colors, as those colors are arranged on the color wheel
**thicket** (n.): a thick, dense growth of shrubs, woody vines, or small trees
**thorn** (n.): a stiff, woody, modified stem with a sharp point
**tin** (n.): stannous chloride, a metallic salt used as a mordant in dyeing
**tomentose** (adj.): woolly, with soft, short, matted or tangled, dense hairs
**toothed** (adj.): having a margin with serrations, dentations, or crenations
**top-dye** (vb.): to apply a dye or coloring substance to a yarn that was previously dyed with some other coloring substance, thereby creating a secondary or tertiary hue
**trailing** (adj.): prostrate or creeping but not rooting
**transverse** (adj.): perpendicular to the long axis of a structure
**tree** (n.): a perennial woody plant of considerable stature at maturity with one main stem (trunk), generally branching well aboveground and exhibiting a well-developed crown
**trigonous** (adj.): having three angles and three sides
**truncate** (adj.): ending abruptly, with the base or apex transversely straight or nearly so, as if cut off
**tube** (n.): a hollow cylindrical structure, such as the constricted basal portion of some fused corollas
**tuber** (n.): a short, thickened branch of subterranean rhizome bearing nodes and buds (for example, a potato)
**tuft** (n.): a dense cluster of organs or structures
**turbinate** (adj.): top-shaped
**umbel** (n.): an indeterminate, flat-topped or hemispheric inflorescence consisting of flowers borne on pedicels arising from a common point of attachment (in simple umbels, pedicels arise from the end of a peduncle; in compound umbels, pedicels arise from the end of rays that arise from the end of a peduncle) (adj. **umbellate** and **umbelliform**)

**uncinate** (adj.): hooked at the tip

**understory** (n.) **1**: a collective term for the plants growing under the canopy of a forest community; **2**: the layer of shade-tolerant trees between the canopy and the ground of a forest community

**undulate** (adj.): having a wavy margin in a vertical plane, alternatively convex and concave; compare with **sinuate**

**urceolate** (adj.): having an urn-shaped or pitcher-like corolla that is globose to cylindrical but contracted at or just below the mouth

**utricle** (n.): a small, thin-walled, one-seeded, bladder-like fruit (for example, saltbush)

**value** (n.): the lightness (high value) or darkness (low value) of a hue

**valvate** (adj.) **1**: meeting at the edges but not overlapping; **2**: opening by valves breaking away from the septum or septa

**vatting** (n.): the process of reducing a dye to a soluble form for application to a fiber, after which the dye is oxidized to the original insoluble pigment

**vein** (n.): an externally visible vascular bundle (primary veins are the largest of a leaf, occurring either singly as a midvein or as a series; secondary veins arise from primary veins)

**vermifuge** (n.): a substance used to kill parasitic worms

**verticil** (n.): a whorl of leaves or flowers (adj. **verticillate**)

**villous** (adj.): shaggy rather than matted, with long, soft, curly hairs

**vine** (n.): a climbing or scrambling plant with elongate, flexible stems that are not self-supporting

**viscid** (adj.): covered with a sticky or gelatinous exudate and gummy to the touch

**viscous** (adj.): having a sticky or glutinous consistency

**washfast** (adj.): stable and resistant to fading and bleeding in the presence of water

**weed** (n.): a plant species that aggressively colonizes disturbed habitats and cultivated lands

**whorl** (n.) **1**: the arrangement of organs, such as leaves, branches, flowers, or fruits, three or more per node; **2**: two or more cyclic groups of sepals, petals, or stamens (adj. **whorled**)

**wing** (n.) **1**: a thin, flat, membranous expansion of a pericarp in a samara; **2**: a lateral petal of a papilionaceous flower; **3**: a cortical eruption of bark forming a ridge on the branches of some woody plants; **4**: a thin, flat, membranous expansion forming a ridge or flange on an organ

**woodland** (n.): a vegetation type dominated by trees whose crowns do not touch; compare with **forest**

# References

Adrosko, R. 1971. *Natural Dyes and Home Dyeing*. New York: Dover.
Bliss, A. 1981. *A Handbook of Dyes from Natural Materials*. New York: Charles Scribner's Sons.
Brummitt, R. K., and C. E. Powell, eds. 1992. *Authors of Plant Names*. Royal Botanic Gardens, Kew.
Buchanan, R. 1995. *A Dyer's Garden*. Loveland, Colorado: Interweave Press.
Buhler, A. 1948. Primitive dyeing methods. *CIBA Review* 6 (68): 2485–2500.
Burrows, G. E., and R. J. Tyrl. 2001. *Toxic Plants of North America*. Ames: Iowa State University Press.
Cannon, J., and M. Cannon. 1994. *Dye Plants and Dyeing*. London: Herbert Press.
Coffey, T. 1993. *The History and Folklore of North American Wildflowers*. New York: Facts on File.
Correll, D. S., and M. C. Johnston. 1970. *Manual of the Vascular Plants of Texas*. Renner: Texas Research Foundation.
Dean, J. 1999. *Wild Color*. New York: Watson-Guptill.
Ellis, A. 1798. *The Country Dyer's Assistant*. Brookfield, Massachusetts: E. Merriam and Company.
Fernald, M. L. 1950. *Gray's Manual of Botany*. 8th ed. New York: American Book Company.
Flora of North America Editorial Committee. 1993–2003. *Flora of North America North of Mexico*. 7 vols. New York: Oxford University Press.
Foster, S., and J. A. Duke. 2000. *A Field Guide to Medicinal Plants and Herbs of Eastern and Central North America*. 2nd ed. New York: Houghton Mifflin.
Gleason, H. A., and A. Cronquist. 1991. *Manual of Vascular Plants of Northeastern United States and Adjacent Canada*. 2nd ed. Bronx: New York Botanical Garden.
Graham, N. A., and R. F. Chandler. 1990. *Podophyllum*. *Canadian Pharmacology Journal* 123: 330–331.

Great Plains Flora Association. 1986. *Flora of the Great Plains*. Lawrence: University of Kansas Press.

Green, C. L. 1995. *Natural Colourants and Dyestuffs: A Review of Production, Markets and Development Potential*. Rome: Food and Agriculture Organization of the United Nations.

Haberly, L. 1957. *Pliny's Natural History*. New York: Ungar Publishing.

Hamel, P., and M. Chiltoskey. 1975. *Cherokee Plants and Their Uses: A 400-Year History*. Sylva, North Carolina: Herald Publishing.

Harbeson, G. 1938. *American Needlework*. New York: Bonanza Books.

Harris, M. 1766. *The Natural System of Colours*. Reprint. New York: Whitney Library of Design, 1963.

Hatch, S. L., and J. Pluhar. 1993. *Texas Range Plants*. College Station: Texas A&M University Press.

Held, S. 1973. *Weaving: A Handbook for Fiber Craftsmen*. New York: Holt, Rinehart and Winston

Huxtable, R. J. 1989. Human health implications of pyrrolizidine alkaloids and herbs containing them. In *Toxicants of Plant Origin*, vol. 1, Alkaloids. Ed. P. R. Cheeke. Boca Raton, Florida: CRC Press.

Johnson, T. 1999. *CRC Ethnobotany Desk Reference*. Boca Raton, Florida: CRC Press.

Kadolph, S. 1997. Commercial dyeing with natural dyes: a controversial response to environmental awareness. Paper presented at the Fifteenth Annual Ars Textrina International Conference on Textiles, June 1997, at Oklahoma State University, Stillwater.

Kierstead, S. 1950. *Natural Dyes*. Boston: Humphries.

Kindscher, K. 1992. *Medicinal Wild Plants of the Prairie: An Ethnobotanical Guide*. Lawrence: University of Kansas Press.

Knaggs, N. 1992. Dyestuffs of the ancients. *American Dyestuff Reporter* 81: 109–111.

Kosnik, M. A., G. M. Diggs Jr., P. A. Redshaw, and B. L. Lipscomb. 1996. Natural hybridization among three sympatric species *Baptisia* (Fabaceae) species in north central Texas. *Sida* 17: 479–500.

Krochmal, A., and C. Krochmal. 1973. *A Guide to the Medicinal Plants of the United States*. New York: Quadrangle.

Lillie, R. 1979. The red dyes used by ancient dyers: their probable identity. *Journal of the Society of Dyers and Colourists* 92: 57–61.

Little, E. L. 1980. *The Audubon Society Field Guide to North American Trees: Eastern Region*. New York: Alfred A. Knopf.

Mahady, G. B., A. B. Schilling, and C. W. Beecher. 1993. *Sanguinaria canadensis* L. (Sanguinarius): in vitro culture and the production of benzophenanthridine alkaloids. In *Medicinal and Aromatic Plants 5, Biotechnology in Agriculture and Forestry*, vol. 24. Ed. Y. P. S. Bajaj. Berlin: Springer-Verlag. 313–328.

Martin, A. C., H. S. Zim, and A. L. Nelson. 1951. *American Wildlife Plants*. New York: Dover.

Miller, J. H., and K. V. Miller. 1999. *Forest Plants of the Southeast and Their Wildlife Uses*. Raleigh, North Carolina: Southern Weed Science Society.

Millsbaugh, C. F. 1892. *American Medicinal Plants*. Reprint. New York: Dover, 1974.

Mitchell, J., and A. Rook. 1979. *Botanical Dermatology: Plants and Plant Products Injurious to the Skin*. Vancouver, British Columbia: Greengrass.

Moerman, D. E. 1986. *Medicinal Plants of Native America*. 2 vols. Research Reports in Ethnobotany, Contribution 2. University of Michigan Museum of Anthropology Technical Reports, Number 19. Ann Arbor.
Moerman, D. E. 1998. *Native American Ethnobotany*. Portland, Oregon: Timber Press.
Morton J. F. 1982. *Plants Poisonous to People in Florida and Other Warm Areas*. Miami, Florida: Morton.
Munsell, A. H. 1905. *A Color Notation*. Boston: Munsell Color Company.
Munsell Color Services. n.d. *Munsell Book of Color: Glossy Finish Collection*. New Windsor, New York: GretagMacbeth.
Nast, J. 1981. Dyes: an Amerindian applied decorative technique. *The Living Museum* 43: 33–35.
Plant Names Project. 1999. *International Plant Names Index*, http://www.ipni.org/index.html.
Preininger, V. 1986. Chemotaxonomy of Papaveraceae and Fumariaceae. In *The Alkaloids*, vol. 29. Ed. A. Brossi. Orlando, Florida: Academic Press. 1–98.
Richards, L. 1992. Acquisition of female apparel in Oklahoma's Indian Territory, 1850–1910. *Home Economics Research Journal* 21: 50–74.
Richards, L. 1994. Folk dyeing with natural materials in Oklahoma's Indian Territory. *Material Culture* 26: 29–47.
Robinson, S. 1969. *A History of Dyed Textiles*. London: Studio Vista.
Sandberg, G. 1989. *Indigo Textiles: Technique and History*. Asheville, North Carolina: Lark Books.
Sandberg, G. 1994. *The Red Dyes: Cochineal, Madder and Murex Purple*. Asheville, North Carolina: Lark Books.
Sheffield, J. 1784. *Observations on the Commerce of the American States*. London: J. Debrett.
Stearns, M. G. 1964. Family dyeing in colonial New England. *Plants and Gardens* 20: 77–82.
Stratton, J. 1981. *Pioneer Women: Voices from the Kansas Frontier*. New York: Simon and Schuster.
Stubbendieck, J., S. L. Hatch, and C. H. Butterfield. 1992. *North American Range Plants*. 4th ed. Lincoln: University of Nebraska Press.
Terrell, E. E. 1996. Revision of *Houstonia* (Rubiaceae-Hedyotideae). *Systematic Botany Monographs* 48: 1–118.
Tull, D. 1987. *A Practical Guide to Edible and Useful Plants*. Austin: Texas Monthly Press.
Turner, H. 1992. Using natural dyes with children. *Journal for Weavers, Spinners and Dyers* 161: 26.
U.S. Department of Agriculture. 2002. *PLANTS Database*, version 3.5, http://plants.usda.gov/. Baton Rouge, Louisiana: National Plant Data Center.
Van Stralen, T. 1993. *Indigo, Madder and Marigold: A Portfolio of Colors from Natural Dyes*. Loveland, Colorado: Interweave Press.
Weigle, P. 1974. *Ancient Dyes for Modern Weavers*. New York: Watson-Guptill.
Weiner, M. A. 1972. *Earth Medicine—Earth Foods: Plant Remedies, Drugs, and Natural Foods of the North American Indians*. New York: Macmillan.
Wilson, K. 1979. *A History of Textiles*. Boulder, Colorado: Westview Press.
Works Progress Administration. 1937. *The Indian-Pioneer Papers*. Norman, Oklahoma: Western History Collection.

# Index